Study Guide to Accompany

PRINCIPLES OF CHEMISTRY
WITH PRACTICAL PERSPECTIVES

Second Edition

Russell S. Drago

Prepared by

ROBERT J. BALAHURA
University of Guelph
Ontario

Allyn and Bacon, Inc.
Boston London Sydney Toronto

ISBN 0-205-05570-2

Second printing . . . July, 1977

To Jeanne

Contents

CONTENTS

Preface

This Study Guide is written to accompany the second edition of "Principles of Chemistry with Practical Perspectives" by Russell S. Drago. Each chapter of the Study Guide corresponds to that in the text and contains three sections.

      A.   Points of Importance
      B.   Types of Problems
      C.   Test Yourself

Section A provides the "tools of the trade" for all chapters. It contains background information, explanations of new terms and difficult concepts, and helpful hints for solving the problems associated with each chapter. Section B categorizes the problems into "types" and gives examples of each type with detailed solutions. A number of the problems from the text without answers (those marked with an asterisk) are also fully worked out in this section. Part C contains a number of new problems for the student to test his or her progress and understanding. Answers to Part C problems and some of the crucial steps to solving these problems are also provided at the back of the Study Guide.

The heavy emphasis on problem solving reflects the authors' feeling that this is the best way to learn. Hopefully this Study Guide will help students to enjoy and understand chemistry as well as to pass their examinations.

                                      Robert J. Balahura

1

Introduction

A. Points of Importance

This chapter reviews many concepts with which you are already familiar and we will only be concerned with a few.

1. Mathematical Operations

Most chemical calculations involve algebraic manipulations. For the general equation

$$x + y = a + b$$

the following operations can be performed:

(i) addition of the same quantity to both sides of the equation

(eg.) $25 + x + y = a + b + 25$

(ii)   subtraction of the same quantity from both sides of the
         equation

         (eg.)   x + y −25 = a + b −25

(iii)   division by the same quantity on both sides of the equation

         (eg.) $\dfrac{(x + y)}{25} = \dfrac{(a + b)}{25}$

(iv)   multiplication by the same quantity on both sides of
         the equation

         (eg.)   25 x (x + y) = 25 x (a + b)

When manipulating equations in any of the above ways you should

(i) perform all operations in brackets first

         (eg.) $\dfrac{25\ (7 + 8)}{5\ (15 - 10)} = \dfrac{25\ (15)}{5\ (5)} = 15.$

(ii) perform necessary addition and subtraction first then
     multiplication and division.

         (eg.) $\dfrac{25 + 30}{11 - 6} = \dfrac{55}{5} = 11.$

A special case of these rules is often encountered in the form of
the equation

$$\frac{a}{b} = \frac{c}{d}$$

Say you wished to evaluate the quantity "d". You would go about this
in the following way:

2

Multiply both sides of the equation by "d"

$$d\,\frac{a}{b} = d\,\frac{c}{d} \quad \therefore \quad \frac{da}{b} = c$$

Divide both sides of the equation by "a".

$$\frac{da}{b} \cdot \frac{1}{a} = c \cdot \frac{1}{a} \quad \therefore \quad \frac{d}{b} = \frac{c}{a}$$

Multiply both sides of the equation by "b".

$$b\,\frac{d}{b} = b\,\frac{c}{a} \quad \therefore \quad d = \frac{bc}{a}$$

Using the same procedures you can arrive at any of the following

relations

$$a = \frac{bc}{d} \qquad\qquad c = \frac{ad}{b}$$

$$b = \frac{ad}{c} \qquad\qquad ad = bc$$

After a little practice you should easily be able to do this in

your head. These manipulations are sometimes referred to as cross-

multiplication.

Since chemists often deal with very small or very large numbers

you will also need to be familiar with the handling of exponential

numbers. A number such as 210.0 can be written as 2.100×10^2 or a

number such as 0.0000031 can be written as 3.1×10^{-6}. Notice that in

obtaining the number 2.100×10^2 we simply moved the decimal place to the

left twice

$$210.0$$

or to obtain the number 3.1×10^{-6} we moved the decimal place to the

right six times

$$0.000003\,1$$

(egs.)   621,000,000 $= 6.21 \times 10^8$

              621,000,000.0 $= 6.210000000 \times 10^8$

              0.00000000612 $= 6.12 \times 10^{-9}$

              13.57 $=$ 1.357×10^1

              9782.31 $=$ 9.78231×10^3

              0.00536 $=$ 5.36×10^{-3}

Remember when multiplying two exponential numbers you must <u>add</u> the exponents and when dividing two exponential numbers you must subtract the exponents.

(egs.) $(1.0 \times 10^{-6}) \times (2.0 \times 10^{-2}) = 2.0 \times 10^{-8}$

$(1.0 \times 10^{-6}) \times (2.0 \times 10^5) = 2.0 \times 10^{-1}$

$(2.0 \times 10^{-6}) \times (9.0 \times 10^5) = 18. \times 10^{-1} = 1.8 \times 10^0 = 1.8$

$(4.0 \times 10^{-6}) \times (4.0 \times 10^{-6}) = 16. \times 10^{-12} = 1.6 \times 10^{-11}$

$(0.40 \times 10^{-6}) \times (1.0 \times 10^{-6}) = 0.40 \times 10^{-12} = 4.0 \times 10^{-13}$

$(6.0 \times 10^{23}) \div (6.0 \times 10^{-5}) = 1.0 \times 10^{28}$

$(1.0 \times 10^{-5}) \div (1.0 \times 10^{-2}) = 1.0 \times 10^{-3}$

$(1.0 \times 10^{-5}) \div (1.0 \times 10^{-8}) = 1.0 \times 10^3$

When adding or subtracting exponential numbers make sure that both numbers are to the <u>same exponent</u>:

(eg.) $1.02 \times 10^{-3} + 3.60 \times 10^{-1} = 0.0102 \times 10^{-1} + 3.60 \times 10^{-1}$

$= 3.61 \times 10^{-1}$

* Remember the correct number of significant figures.

See section 1-4 in text.

2. Problem Solving

All problems in this guide will be solved using the <u>dimensional analysis</u> or <u>conversion factor</u> approach. Let's illustrate this method with the following examples.

(i) A liquid sample has a mass of 15.9 g and a volume of 12.6 ml. Calculate its density in lbs ft^{-3}.

In this problem you are given the required information but in the wrong units. You need to convert grams to pounds and milliliters to cubic feet. You require the conversion factors given below to carry out the required changes:

$$454 \text{ g} = 1 \text{ lb}$$
$$1 \text{ ml} = 1 \text{ cm}^3$$
$$2.54 \text{ cm} = 1 \text{ in}$$
$$12 \text{ in} = 1 \text{ ft}$$

Let's worry about the change from grams to pounds first. The conversion factor 454 g = 1 lb can be written as

$$\frac{454 \text{ g}}{1 \text{ lb}} \text{ or } \frac{1 \text{ lb}}{454 \text{ g}}.$$

Both of these quantities are equal to unity since the numerator and denominator are equal (454g = 1 lb). Now, to convert 15.9g to lbs we simply multiply by the appropriate conversion factor:

$$15.9 \text{ g} \times \boxed{\frac{1 \text{ lb}}{454 \text{ g}}} = 0.0350 \text{ lb} = 3.50 \times 10^{-2} \text{ lb}$$

Notice that the units of g cancel out leaving the required units of lbs. If you had inadvertently used the inverted form of the conversion factor,

$\frac{454 \text{ g}}{1\text{b}}$, the units would have been $\frac{\text{g}^2}{1\text{b}}$ which you would immediately recognize to be incorrect.

The conversion of ml to ft^3 can also be carried out in a similar manner.

$$12.6 \text{ ml} \times \frac{1 \text{ cm}^3}{\text{ml}} \times \frac{(1 \text{ in})^3}{(2.54 \text{ cm})^3} \times \frac{(1 \text{ ft})^3}{(12 \text{ in})^3} = 4.45 \times 10^{-4} \text{ ft}^3$$

Notice that we had to "cube" the last two conversion factors in order to get our units to cancel. Now we can obtain the correct density:

$$\frac{3.50 \times 10^{-2} \text{ lb}}{4.45 \times 10^{-4} \text{ ft}^3} = 0.787 \times 10^2 \text{ lb ft}^{-3}$$

We could have carried out both steps together by writing

$$\frac{15.9 \text{ g}}{12.6 \text{ ml}} \times \frac{1 \text{ lb}}{454 \text{ g}} \times \frac{1 \text{ ml}}{1 \text{ cm}^3} \times \frac{(2.54 \text{ cm})^3}{(\text{ in})^3} \times \frac{(12 \text{ in})^3}{(\text{ ft})^3} = 0.787 \times 10^2 \frac{\text{lb}}{\text{ft}^3}$$

(ii) Given that a unit called a mole* of Hg weighs 200.59g, how many grams of Hg are there in 0.00631 moles of mercury?

Write down what we are given:

0.00631 moles Hg or 6.31×10^{-3} moles Hg.

What do we know about the relation between the number of moles of Hg and the number of grams of Hg?

---

* The mole is defined in section 3 following and in section 1-8 of the text.

6

Our conversion factor is

$$\boxed{\dfrac{200.59 \text{ g Hg}}{1 \text{ mole Hg}}}$$

In other words, this relation says that there are 200.59 g Hg
per mole of Hg.

We can now solve the problem

$$6.31 \times 10^{-3} \text{ moles Hg} \times \frac{200.59 \text{ g Hg}}{\text{mole Hg}} = 1.27 \text{ g Hg.}$$

Notice that the "moles of Hg" cancel out leaving the correct
units, "g Hg". Again, you may ask the question "how do I know
whether to use $\frac{200.59 \text{ g Hg}}{1 \text{ mole Hg}}$ or $\frac{1 \text{ mole Hg}}{200.59 \text{ g Hg}}$"? The units in the
final answer will immediately tell you if you have chosen the
correct conversion factor. For this reason you are urged to
include units in all steps of your calculations.

Some of you may have recognized that dimensional analysis is a short
form of the "statements-ratio" method of solving problems. i.e.

1 mole Hg contains 200.59 g Hg

6.31×10^{-3} moles Hg contains x g Hg

$\therefore \dfrac{1 \text{ mole Hg}}{6.31 \times 10^{-3} \text{ mole Hg}} = \dfrac{200.59 \text{ g Hg}}{\text{x g Hg}}$

Solving for x, we obtain

$$x = 6.31 \times 10^{-3} \text{ mole Hg} \times \frac{200.59 \text{ g Hg}}{1 \text{ mole Hg}}$$

$$= 1.27 \text{ g Hg.}$$

This is exactly the same form of the equation as we obtained previously. At the beginning you may need to use the "statements-ratio"
method in thinking out the problem until you become more familiar
with dimensional analysis.

3. The MOLE and Atomic Weights, Gram Atomic Weights, Gram Atoms,

    Molecular Weights and Gram Molecular Weights.

Central to the discussion of atoms and atomic weights is the
concept of the mole. A mole of atoms, or of molecules, is the amount
whose mass in grams is equal to the atomic weight, or molecular
weight.* From the definition we can write the equation

$$1 \text{ mole} = AW(g) \text{ or } MW(g)$$

and the conversion factors

$$\frac{AW(g)}{1 \text{ mole}} \quad \text{or} \quad \frac{MW(g)}{1 \text{ mole}} .$$

However, a mole of atoms, or of molecules, must always consist of the
same number of atoms or molecules, and this is Avogadro's number
6.023×10^{23}. From this, we get another useful conversion factor

$$\frac{6.023 \times 10^{23} \text{ anything}}{1 \text{ mole anything}} .$$

_____

* The terms gram-atom, gram atomic weight, gram-mole, and gram
molecular weight are all equivalent to this definition of the mole.
Most chemists, however, use the simpler term "mole" and apply it to
both atoms and molecules.

_____

It can be very important to know the relationship between the atomic weight of an element as given in a periodic table, for instance, and the masses of individual atoms. Although the atomic weight of oxygen is given as 15. 9994, no individual oxygen atom has this mass. This is because oxygen has three naturally occurring isotopes, $^{16}_{8}O$, $^{17}_{8}O$, and $^{18}_{8}O$, with percent abundances of 99.759, 0.037, and 0.204, respectively. The masses of individual atoms of these isotopes are 15.9949, 16.9991, and 17.9992, respectively. The value of 15.9994 given for the atomic weight is the weighted average of these isotopic masses. Since the abundances quoted are atom percentages, not weight percentages, one mole of oxygen atoms will contain 0.99759 mole of $^{16}_{8}O$ atoms, 0.00037 mole $^{17}_{8}O$ atoms, and 0.00204 mole $^{18}_{8}O$ atoms. The masses of these amounts of the individual isotopes can be added to give the mass of a mole of naturally distributed oxygen atoms:

$$0.99759 \text{ mole } {}^{16}_{8}O \times 15.9949 \text{ g/mole} = 15.9564 \text{ g } {}^{16}_{8}O$$

$$0.00037 \text{ mole } {}^{17}_{8}O \times 16.9991 \text{ g/mole} = 0.0063 \text{ g } {}^{17}_{8}O$$

$$0.00204 \text{ mole } {}^{18}_{8}O \times 17.9992 \text{ g/mole} = 0.0367 \text{ g } {}^{18}_{8}O$$

$$\overline{1.00000 \text{ mole } O = 15.9994 \text{ g } O}$$

The average atomic weight of O is, therefore, 15.9994, since one mole of O has a mass of 15.9994 g. This relationship is illustrated in reverse in Example 2 in the next section.

B. Types of Problems

There are only two types of problem that you must recognize:

(1) those involving some combination of <u>moles-molecules-grams</u> and

(2) those involving <u>average atomic weight</u> and <u>percent natural abundance</u>.

Examples

1.(a) Calculate the mass of oxygen in 2.00 g of sodium dichromate dihydrate, $Na_2Cr_2O_7 \cdot 2H_2O$. The gram-molecular-weight (GMW) of $Na_2Cr_2O_7 \cdot 2H_2O$ is 298.00 g.

There are a number of relationships which will enable you to solve this problem and <u>they all use the idea that 1 mole or 1 GMW of $Na_2Cr_2O_7 \cdot 2H_2O$ weighs 298.00 g.</u>

The weight of oxygen atoms in 1 mole of this compound is

$$9 \text{ moles oxygen atoms} \times \frac{15.999 \text{ g O}}{\text{mole oxygen atom}} = 143.99 \text{g O}$$

∴ we know that there are 143.99 g O per mole of $Na_2Cr_2O_7 \cdot 2H_2O$ or per 298.00g $Na_2Cr_2O_7 \cdot 2H_2O$.

∴ our relationship is $\dfrac{143.99 \text{ g O}}{298.00 \text{ g } Na_2Cr_2O_7 \cdot 2H_2O}$

Finally we have $2.00 \text{ g } Na_2Cr_2O_7 \cdot 2H_2O \times \dfrac{143.99 \text{ g O}}{298.00 \text{ g } Na_2Cr_2O_7 \cdot 2H_2O}$

$$= 0.966 \text{ g O}$$

(b) Calculate the mass of 3.00×10^{18} molecules of ethylene glycol, $C_2H_6O_2$.

This time we are relating molecules and grams. This again involves the mole. We know that 1 mole of ethylene glycol must contain 6.023×10^{23} molecules. We also know that 1 mole of ethylene glycol weighs 62.07 g.

$\therefore$ our relationship is $\dfrac{62.07 \text{ g}}{6.023 \times 10^{23} \text{ molecules}}$

and 3.00×10^{18} molecules $\times \dfrac{62.07 \text{ g}}{6.023 \times 10^{23} \text{ molecules}}$

$$= 3.09 \times 10^{-4} \text{ g}.$$

(c) Calculate the number of moles of carbon in 8.00 g of sucrose, $C_{12}H_{22}O_{11}$.

8.00 g sucrose $\times \dfrac{1 \text{ mole sucrose}}{342.30 \text{ g sucrose}} \times \dfrac{12 \text{ moles carbon}}{1 \text{ mole sucrose}}$

$$= 0.280 \text{ moles of carbon}.$$

If the question had asked for the number of atoms of carbon we would have used a third factor

i.e. $\dfrac{6.023 \times 10^{23} \text{ atoms carbon}}{1 \text{ mole carbon}}$ Note that in multi-plying 8.00 g sucrose by $\dfrac{1 \text{ mole sucrose}}{342.30 \text{ g sucrose}}$ we were deter-mining the number of moles in a given weight of sucrose. This is a very common calculation in chemistry.

2. Copper consists of two naturally occurring isotopes of masses 62.9298 and 64.9278. Using the atomic weight of Cu (from a periodic table) of 63.54 estimate the abundances of these isotopes.

The AW of Cu, 63.54, represents a weighted average of the two isotopes. This weighted average is obtained by estimating the contribution to the "average weight" of each isotope i.e.

(% abundance) x 62.9298 + (% abundance) x 64.9278.

Let the fraction of the isotope of mass 62.9298 be x. Since the percent abundances must add up to 100 then the fraction of the isotope of mass 64.9278 must be 1-x.

∴ we have 63.54 = (x)(62.9298) + (1-x)(64.9278)

$$64.9278x - 62.9298x = 64.9278 - 63.54$$

$$1.9980x = 1.39$$

$$x = 0.696$$

∴ the % abundance of isotope 63 is 69.6% and the % abundance of isotope 65 is 30.4%.

<u>Test Yourself</u>

* Make sure you can recognize the types of problems.

1. (i)   Calculate the mass in grams of one atom of $^{238}_{92}U$.

   (ii)   Calculate the mass in grams of one atom of $^{7}_{3}Li$.

   (iii) Calculate the mass in grams of one molecule of the
          insecticide parathion, $C_{10}H_{14}NO_5PS$.

   (iv)   Calculate the number of molecules of cholesterol, $C_{27}H_{46}O$,
          in 25.0 g cholesterol.

   (v)   Calculate the number of atoms of carbon in 28.0 g of
          tetraethyl lead, $(C_2H_5)_4Pb$.

   (vi.) Calculate the mass of carbon in 88.0 g of dimethyl mercury,
          $(CH_3)_2Hg$.

   (vii) Calculate the number of moles of the insecticide DDT,
          $C_{14}H_9Cl_5$, in a 454 g sample.

2.   Magnesium exists as three isotopes of atomic mass 23.98504, 24.98584,
     and 25.98259 with percent natural abundances 78.70, 10.13, and 11.17%
     respectively. Calculate the average atomic mass of naturally occurring
     magnesium.

3.   The element hydrogen consists of two naturally occurring isotopes
     with masses 1.0078 and 2.0141. Using the average atomic weight of
     hydrogen of 1.0080 estimate the abundances of these isotopes.
     Write the symbol and common name for each of these isotopes.

4. Complete the following table:

Species	number of protons	number of electrons	number of neutrons
$^{32}_{16}S$	16	16	16
$^{79}_{34}Se^{2-}$	34	36	45
$^{56}_{26}Fe^{3+}$	26	23	30
$^{40}_{20}Ca^{2+}$	20	18	20
$^{27}_{13}Al^{3+}$	13	10	14
$^{19}_{9}F^{1-}$	9	10	10

14

2

Tools of the Trade

A. Points of Importance

1. Chemical Formulas and Moles

The molecular formula of nicotine, a tobacco alkaloid, is $C_{10}H_{14}N_2$. A mole of nicotine contains 6.023×10^{23} molecules of nicotine. Each molecule is made up of 10 atoms of carbon, 14 atoms of hydrogen and 2 atoms of nitrogen. Thus each mole of nicotine contains 10 moles of carbon atoms, 14 moles of hydrogen atoms and 2 moles of nitrogen atoms. The gram-molecular-weight (GMW) of nicotine can be obtained by summing the gram-atomic-weights of all the atoms making up the molecule:

$$10(12.00) + 14(1.008) + 2(14.01) = 162.2 \text{ g mole}^{-1}$$

Thus the relative number of atoms in a molecule is given by the relative number of moles of atoms for a given sample. Because of the relation of the mole to GMW we can easily calculate the percentage by weight of each element in a pure compound if we know the molecular formula. For example nicotine has the following composition:

$$\%C = \frac{\text{total mass carbon in sample}}{\text{total mass of sample}} \times 100$$

As usual, a convenient choice is 1 mole of nicotine.

$$\therefore \ \%C = \frac{120.1 \text{ g C}}{162.2 \text{ g nicotine}} \times 100 = 74.04\% \text{ C}$$

Similarly,

$$\%H = \frac{14.11 \text{ g H}}{162.2 \text{ g nicotine}} \times 100 = 8.70\% \text{ H}$$

$$\%N = \frac{28.02 \text{ g N}}{162.2 \text{ g nicotine}} \times 100 = 17.27\% \text{ N.}$$

These values are also the percentages in one molecule of nicotine. If we have some means of determining the relative number of moles of the various atoms in a pure unknown compound we can specify the composition of the substance. A useful way of determining the relative number of moles of the atoms in an unknown is to convert all of one of the elements in the unknown to a known compound. For example, the chloride present in an unknown may be determined by reacting a known weight of the unknown with an excess of silver ion to form silver chloride, AgCl. From the experimentally determined weight of AgCl formed one can determine the amount of Cl in the unknown compound. We will illustrate this general principle in section B by means of an example problem.

Note that analysis gives the simplest whole number ratio of atoms in a compound and this is called the simplest or empirical formula. If the

16

experimentally determined molecular weight is the same as the weight
of the empirical formula, then the empirical formula must also be the
molecular formula. In the case of nicotine the empirical formula is
found to be C_5H_7N which has a weight of 81.1. The experimentally
determined gram molecular weight is however 162.2 g mole^{-1}. The molecular
formula is twice the empirical formula or $C_{10}H_{14}N_2$. This agrees with the
actual structure of nicotine which involves the combination of the atoms
in $C_{10}H_{14}N_2$ to form one molecule.

2. Chemical Equations

Chemical equations are a short-hand notation for conveying information about chemical reactions. The formulas of the reactants tell us
what compounds react and the formulas of the products tell us what
compounds are formed. The equation also tells us how much of the reactants
will be required to produce a given amount of product. For example, the
synthesis of DDT is represented by the following equation

$$2C_6H_5Cl + C_2H_3O_2Cl_3 \rightarrow C_{14}H_9Cl_5 + 2H_2O.$$
$$\text{DDT}$$

The equation tells us that if we have the right amounts of C_6H_5Cl and
$C_2H_3O_2Cl_3$ (the stoichiometric amounts) we can produce exactly 1 mole of

DDT and 2 moles of water. In other words the equation tells us the maximum amount of DDT that can be produced. <u>How much DDT can be produced from 1.5 moles of C_6H_5Cl and an excess of $C_2H_3O_2Cl_3$?</u>

The first thing you should realize is that you want a relation between moles of C_6H_5Cl and moles of DDT. Since you are given a chemical equation you have the relation

$$\boxed{2 \text{ moles } C_6H_5Cl = 1 \text{ mole DDT}}$$

Therefore you know

$$2 \text{ moles } C_6H_5Cl \text{ produce } 1 \text{ mole DDT}$$

and $\qquad$ $1.5 \text{ moles } C_6H_5Cl \text{ produces} \qquad ? \text{ moles DDT}$

$$\therefore \qquad \frac{2 \text{ moles } C_6H_5Cl}{1.5 \text{ moles } C_6H_5Cl} = \frac{1 \text{ mole DDT}}{? \text{ moles DDT}}$$

$$? = 1.5 \text{ moles } C_6H_5Cl \times \frac{1 \text{ mole DDT}}{2 \text{ mole } C_6H_5Cl} = 0.75 \text{ mole DDT}$$

We could have used a ratio directly to solve the problem as below

$$1.5 \text{ moles } C_6H_5Cl \times \boxed{\frac{1 \text{ mole DDT}}{2 \text{ mole } C_6H_5Cl}} = 0.75 \text{ mole DDT}.$$

Again you would have detected incorrect units if you had used the ratio $\boxed{\dfrac{2 \text{ mole } C_6H_5Cl}{1 \text{ mole DDT}}}$ by mistake.

All the problems involving equations can be solved using the above approach. The only difference is that you are usually given information in grams and asked for amounts in grams. You need only be able to convert everything to moles at the beginning to obtain an answer in moles and

18

finally to convert to grams if necessary. Thus this one approach is
sufficient for all problems.

Let's now go one step further. In the previous example we used an
excess of $C_2H_3O_2Cl_3$ so we knew that we had enough $C_2H_3O_2Cl_3$ to react
with the 1.5 moles of C_6H_5Cl and some unspecified amount would be
left over after all of the 1.5 moles of C_6H_5Cl was used up. So we
really only <u>specified</u> (gave a number for) the amount of one of the
reactants. Say now that we wish to know <u>how much DDT will be produced</u>
<u>by the reaction of 2 moles of C_6H_5Cl with 2 moles of $C_2H_3O_2Cl_3$</u>. We
have <u>specified</u> the amounts of each of the reactants so which do we
use for our calculation? We must find out which reactant is in excess
because the amount of DDT produced will be <u>limited</u> by the reactant in
deficit. (Just as in the first example we did the calculation with
the reactant in deficit). In this example it is easy to see that the
reactant in deficit is C_6H_5Cl (we have 2 moles). Therefore once we
have used up the 2 moles of C_6H_5Cl the reaction must stop leaving
behind 1 mole of unreacted $C_2H_3O_2Cl_3$. Since $C_2H_3O_2Cl_3$ is in excess we
must use the other reagent, that in deficit, in our calculations:

$$2 \text{ moles } C_6H_5Cl \times \frac{1 \text{ mole DDT}}{2 \text{ moles } C_6H_5Cl} = 1 \text{ mole DDT}$$

$\uparrow$

from the chemical equation

19

To find out which reagent was limiting or in deficit we had to look at
the coefficients in the equation. In this example it was easy since the
coefficients of the reactants were 2 and 1. These coefficients can be more
complex depending on the reaction type and the number of reactants. A
useful way of checking your logic is to divide the number of moles of each
reactant available by its coefficient in the equation. The smallest number
obtained will then tell you the limiting reagent. This procedure simply
brings in the stoichiometry of the reaction mathematically. We always use
the limiting reagent in our calculations when given more than one piece of
information about the reactants. This example tells us that if we have 2
moles of C_6H_5Cl no matter how much more than 1 mole of $C_2H_3O_2Cl_3$ we use
we can still only produce 1 mole DDT. From the combinations of reagents
given below you should be able to predict the yields of DDT and water.

$2C_6H_5Cl + C_2H_3O_2Cl_3 \longrightarrow C_{14}H_9Cl_5 + 2H_2O$				leftover
2	2	1	2	1 mole $C_2H_3O_2Cl_3$
$\frac{5}{2}$ 2.5	1	1	2	
1	1			
2	0.5			
4	2			

Remember the fundamental quantity in these calculations is the mole. In all
of these calculations we are determining the theoretical yield of products
i.e. the maximum amount of products that can be produced. In order to calculate a percent yield we must be given the actual yield.

$$\% \text{ yield} = \frac{\text{actual yield}}{\text{theoretical yield}} \times 100$$

20

B. Types of Problems

There are five types of problems for you to recognize, all involving
as the major step, the manipulation of moles as shown in part A.

   (1)   Analysis to obtain percentages. There are many variations
on this type of problem but the same principles are
involved - relating moles of a known species to an
unknown species.

   (2)   Determination of the empirical or simplest formula.

   (3)   Simple stoichiometry and extensions.

   (4)   Limiting reagent.

   (5)   Percent yield.

Examples

(1)   A compound gives the following analysis:

% C = 36.98, % H = 2.22, % N = 18.50, % O = 42.27.

Determine the empirical formula. Assume 100 g of compound and
calculate the relative number of moles of each atom.

$$\text{moles C} = \frac{36.98 \text{ g}}{12.00 \frac{\text{g}}{\text{mole}}} = 3.082$$

$$\text{moles H} = \frac{2.22 \text{ g}}{1.007 \frac{\text{g}}{\text{mole}}} = 2.20$$

$$\text{moles N} = \frac{18.50 \text{ g}}{14.007 \frac{\text{g}}{\text{mole}}} = 1.321$$

$$\text{moles O} = \frac{42.27 \text{ g}}{16.00 \frac{\text{g}}{\text{mole}}} = 2.642$$

21

Divide each of the above by the smallest number of moles to

obtain whole numbers (÷ by 1.321):

$$\therefore \quad \frac{3.082 \text{ moles C}}{1.321 \text{ moles N}} = 2.333 \frac{\text{moles C}}{\text{mole N}}$$

$$\frac{2.20 \text{ moles H}}{1.321 \text{ moles N}} = 1.67 \frac{\text{moles H}}{\text{mole N}}$$

$$\frac{1.321 \text{ moles N}}{1.321 \text{ moles N}} = 1.000$$

$$\frac{2.642 \text{ moles O}}{1.321 \text{ moles N}} = 2.000 \frac{\text{moles O}}{\text{mole N}}$$

Multiply by 3 to obtain whole numbers.

$\therefore$ we have 7 moles C, 5 moles H, 3 moles N, and 6 moles O.

Since the relative number of moles must also be the relative

number of atoms, the empirical formula is $C_7H_5N_3O_6$. This

happens to be the molecular formula for TNT.

(2) (i) An analyst is given a white powder suspected of being a

dangerous drug and asked to determine its formula. He finds that

the powder contains only carbon, hydrogen, and oxygen and decides

to analyse for carbon and hydrogen. He burns 3.00 mg of powder

and obtains 6.60 mg of carbon dioxide and 1.20 mg of H_2O.

From this information he can calculate the percent C, H, and O

and thus determine the simplest or empirical formula. First

he calculates how much carbon is contained in 6.60 mg CO_2.

Since all of this carbon came from the 3.00 mg sample of powder burned he can calculate the percent carbon in the white powder. Since every mole of CO_2 weighs 44.00 g and contains 12.00 gC the required ratio is $\boxed{\dfrac{12.00 \text{ gC}}{44.00 \text{ gCO}_2}}$. The 6.60 mg CO_2 contains

$$6.60 \times 10^{-3} \text{gCO}_2 \times \frac{12.00 \text{ gC}}{44.00 \text{ gCO}_2} = 1.80 \times 10^{-3} \text{ gC}.$$

Since all of this carbon came from the original sample

$$\% \text{ C in sample} = \frac{1.80 \times 10^{-3} \text{ gC}}{3.00 \times 10^{-3} \text{ g sample}} \times 100 = 60.00\%$$

Now for hydrogen he carries out a similar calculation using the amount of H_2O formed:

since every mole of H_2O weighs 18.01 g and contains 2.014 g H the required ratio is $\boxed{\dfrac{2.014 \text{ g H}}{18.01 \text{ g H}_2O}}$ and the 1.20 mg H_2O con-

tains $1.20 \times 10^{-3} \text{ g H}_2O \times \dfrac{2.014 \text{ g H}}{18.01 \text{ g H}_2O} = 0.134 \times 10^{-3} \text{ g H}.$

Since all of this hydrogen came from the original sample

$$\% \text{ H in sample} = \frac{0.134 \times 10^{-3} \text{ g H}}{3.00 \times 10^{-3} \text{ g sample}} \times 100 = 4.46\% \text{ H}.$$

Since the unknown compound only contains C, H, and O we can obtain the % O since the % C plus % H plus % O must add up to 100%.

$$\therefore \% \text{ O} = 100.00 - 60.00 - 4.46 = 35.54\%.$$

Now we can calculate the empirical formula following the rules given in the text.

First, assume 100 g compound since we have percentages.

Second, calculate the number of moles of each element.

23

<u>Third</u>, reduce the numbers obtained to whole number ratios.

$$\text{moles carbon} = \frac{60.00 \text{ g}}{12.0 \frac{\text{g}}{\text{mole}}} = 5.00 \text{ moles}$$

$$\text{moles hydrogen} = \frac{4.46 \text{ g}}{1.007 \frac{\text{g}}{\text{mole}}} = 4.43 \text{ moles}$$

$$\text{moles oxygen} = \frac{35.54 \text{ g}}{16.0 \frac{\text{g}}{\text{mole}}} = 2.22 \text{ moles}$$

Reducing to whole numbers is accomplished by dividing the number
of moles of each atom by the smallest number of moles (always).
In this case divide by 2.22. This gives

$$\text{carbon:} \quad \frac{5.00 \text{ moles C}}{2.22 \text{ moles O}} = 2.25 \frac{\text{moles C}}{\text{mole O}}$$

$$\text{hydrogen:} \quad \frac{4.43 \text{ moles H}}{2.22 \text{ moles O}} = 2.00 \frac{\text{moles H}}{\text{mole O}}$$

$$\text{oxygen:} \quad \frac{2.22 \text{ moles O}}{2.22 \text{ moles O}} = 1.00$$

Multiplication by 4 gives 9 moles of C, 8 moles of H, and
4 moles of O. Since the relative number of moles must also be
the relative number of atoms the empirical formula is $C_9H_8O_4$.
A separate experiment to determine the gram molecular weight
gives a value of 182 g mole^{-1}. Since the empirical formula weight
is also 182 the molecular formula is $C_9H_8O_4$. This is low for
some of the more common mind-bending drugs and running
the mass spectrum confirms ordinary aspirin. The molecular
formula for aspirin is indeed $C_9H_8O_4$.

> Note that by forming the known compounds, H_2O and CO_2, the analyst was able to calculate the percentages of C and H which eventually lead to the empirical formula.

(ii) A 7.275 g silver ring from Mexico (stamped .925) is dissolved in nitric acid. When excess sodium chloride is added to the solution all the silver is precipitated as AgCl. The AgCl precipitate is found to weigh 9.000 g. What is the percentage of silver in the ring?

> In this case we are forming the known compound AgCl. From the weight of AgCl obtained we can calculate the amount of silver. Since all of this silver came from the ring we can calculate the percent silver in the ring.

For the known compound we obtain our ratio $\boxed{\dfrac{107.87 \text{ g Ag}}{143.32 \text{ g AgCl}}}$. Now, how much Ag is there in 9.000 g AgCl?

$$9.000 \text{ g AgCl} \times \frac{107.87 \text{ g Ag}}{143.32 \text{ g AgCl}} = 6.774 \text{ g Ag}$$

All of this Ag came from the ring, therefore

$$\% \text{ Ag} = \frac{6.774 \text{ g Ag}}{7.275 \text{ g ring}} \times 100 = 93.11\%.$$

Again note that <u>the key to the analysis is knowing the amount of Ag in AgCl</u>. This is a basic principle for problems of this type.

(3) Nitroglycerine, $C_3H_5(NO_3)_3$ can be produced according to the equation

$$C_3H_5(OH)_3 + 3HNO_3 \rightarrow C_3H_5(NO_3)_3 + 3H_2O.$$
glycerine

How many grams of glycerine are required to produce 25.0 g of nitroglycerine.

First, in any problem <u>always</u> convert to moles

$$25.0 \text{ g nitroglycerine} \times \frac{1 \text{ mole nitroglycerine}}{227.05 \text{ g nitroglycerine}} = 0.110 \text{ mole nitroglycerine.}$$

Since we do this so often we no doubt have memorized the formula

$moles = \dfrac{grams}{GMW}$ and $\therefore$ from now on we will simply use this with the appropriate units:

i.e. $\dfrac{25.0 \text{ g}}{227.05 \dfrac{\text{g}}{\text{mole}}} = 0.110 \text{ mole.}$

We now have the moles of nitroglycerine required and wish to obtain, from the equation, the factor which will give us moles of glycerine required. The factor is $\dfrac{1 \text{ mole glycerine}}{1 \text{ mole nitroglycerine}}$

$\therefore$ 0.110 mole nitroglycerine $\times \dfrac{1 \text{ mole glycerine}}{1 \text{ mole nitroglycerine}} = 0.110$ moles glycerine.

$\therefore$ # g glycerine = moles x GMW

$= 0.110 \text{ moles} \times \dfrac{92.10 \text{ g}}{\text{mole}}$

$= 10.1 \text{ g glycerine.}$

The key to this problem and its variations is using the equation to relate the moles of the known to the required compound.

(4) Freon 12, a gas used as a refrigerant, can be prepared by reaction
of antimony trifluoride and carbon tetrachloride according to the
equation:

$$2SbF_3 + 3CCl_4 \longrightarrow 3CCl_2F_2 + 2SbCl_3.$$
$$\text{Freon 12}$$

Calculate the maximum yield of Freon 12 which can be prepared from
50.0 g SbF_3 and 60.0g CCl_4. Since we are given two pieces of information about the reactants we immediately realize we have a
limiting reagent problem. Always follow the same procedure to
solve these problems:

<u>Determine the number of moles of each reagent</u>.

$$\dfrac{50.0 \text{g } SbF_3}{\dfrac{178.74\text{g } SbF_3}{\text{mole } SbF_3}} = 0.280 \text{ moles } SbF_3$$

$$\dfrac{60.0 \text{g } CCl_4}{\dfrac{153.81\text{g } CCl_4}{\text{mole } CCl_4}} = 0.390 \text{ moles } CCl_4$$

<u>Divide by the coefficients in the equation</u>.

$$\dfrac{0.280 \text{ moles } SbF_3}{2} = 0.140 \text{ moles } SbF_3$$

$$\dfrac{0.390 \text{ moles } CCl_4}{3} = 0.130 \text{ moles } CCl_4$$

The limiting reagent is the reactant in a molar deficit <u>after</u>
<u>taking the stoichiometry into account</u>.

27

Therefore CCl_4 is the limiting reagent.

Thus we only work with the moles of CCl_4

i.e. 0.390 moles CCl_4.

From the equation we obtain our conversion factor $\boxed{\dfrac{3 \text{ moles Freon 12}}{3 \text{ moles } CCl_4}}$

$\therefore$ 0.390 moles CCl_4 x $\dfrac{3 \text{ moles Freon 12}}{3 \text{ moles } CCl_4}$ = 0.390 moles Freon 12.

$\therefore$ g Freon 12 = moles x GMW

= 0.390 moles x 120.90 $\dfrac{g}{mole}$

= 47.2 g.

(5) A percent yield problem is exactly the same as the previous two types except that an additional piece of information must be given – that of the <u>actual amount of material obtained experimentally</u>. In the previous two types we were calculating theoretical yield. Consider the equation

$$2SbF_3 + 3CCl_4 \longrightarrow 3CCl_2F_2 + 2SbCl_3$$

Calculate the percent yield of Freon 12 if 36.8 g of Freon 12 were isolated from the reaction of 50.0g SbF_3 and 60.0g CCl_4. First we must calculate the theoretical yield. From the previous problem this was 47.2g.

$\therefore$ $\boxed{\% \text{ yield} = \dfrac{\text{actual yield}}{\text{theoretical yield}} \text{ x } 100}$ = $\dfrac{36.8g}{47.2g}$ x 100 = 78.0%.

C. Test Yourself

* Can you recognize the types of problems? Categorize the problems
  in the textbook at the end of chapter 2.

1. Calculate the empirical formula of a compound with the composition
   24.77% Co, 29.80% Cl, 40.35% O, and 5.08% H.

2. Calculate the empirical formula of a compound with the composition
   26.58% K, 35.35% Cr, and 38.07% O.

3. You are asked to analyse a sample suspected of being the tranquilizer
   meprobamate sold under the name Equanil. Meprobamate contains C,H,N,
   and O. Burning 7.50 mg of the compound produced 13.6 mg CO_2 and
   5.57 mg H_2O. A separate analysis for N using 9.50 mg of sample produced 1.48 mg of NH_3. Calculate the empirical formula for meprobamate.

4. How many moles of the pollutant SO_2 are produced from 5.00×10^3
   moles of the ore $CuFeS_2$?

$$2CuFeS_2 + 5O_2 \longrightarrow 2Cu + 2FeO + 4SO_2.$$

5. How many moles of the pollutant SO_2 are produced from 5.00×10^3 moles
   of the ore Cu_2S?

$$Cu_2S + O_2 \longrightarrow 2Cu + SO_2.$$

6. How many grams of P_4 can be produced from 25.0g SiO_2 according to the
   reaction

$$2Ca_3(PO_4)_2 + 6SiO_2 + 5C \longrightarrow P_4 + 6CaSiO_3 + 5CO_2?$$

7. A mixture of KCl and $BaCl_2$ weighing 0.525g was dissolved in water and excess sulphuric acid added to precipitate the barium as $BaSO_4$. If the weight of $BaSO_4$ was 0.1632g what percent of the mixture was $BaCl_2$?

8. For the reaction

$$2Ca_3(PO_4)_2 + 6SiO_2 + 5\ C \longrightarrow P_4 + 6CaSiO_3 + 5CO_2$$

   (i) What is the limiting reagent if 1 mole of each reactant is used?

   (ii) What is the limiting reagent if 100.0g of each reactant are used?

9. For the reaction

$$2KMnO_4 + 5H_2O_2 + 6HCl \longrightarrow 2MnCl_2 + 5O_2 + 8H_2O + 2KCl$$

   (i) How many grams of $MnCl_2$ are produced from reaction of 20.0 g $KMnO_4$, 10.0 g H_2O_2, and 1.00×10^2 g HCl?

   (ii) If the actual yield is 9.82 g $MnCl_2$ what is the percent yield?

10. Consider the following reactions

$$3A + B \longrightarrow 5C$$
$$2C + D \longrightarrow 3E$$
$$E + F \longrightarrow 6P$$

How many moles of P can be produced from 1 mole of A?

3

Solutions

A. Points of Importance

1. Expressing the concentration of a Solution.

Three of the most common methods for expressing the concentration of a solution (amount per "something") are weight percent, molarity, and parts per million. Their definitions are summarized below:

$$\text{weight percent solute} = \frac{\text{weight solute}}{\text{weight solute + weight solvent}} \times 100$$

$$= \frac{\text{weight solute}}{\text{total weight}} \times 100.$$

$$\text{molarity} = \frac{\text{number of moles solute}}{\text{number of liters solution}} = M$$

$$\text{parts per million} = \frac{\text{amount}}{1 \times 10^6 \text{ units}} = \frac{\text{grams}}{1 \times 10^6 \text{ grams}}.$$

In addition to the above a useful expression of concentration is the mole fraction:

$$\text{mole fraction solute} = \frac{\text{number of moles solute}}{\text{number of moles solute} + \text{number of moles solvent}}$$

$$= \frac{\text{number of moles solute}}{\text{total number of moles}} .$$

By far the most important of the above is <u>molarity</u>.

2. <u>Reactions in Solution</u>.

Reactions of liquid reagents of given concentration are handled in exactly the same way as the simple stoichiometry problems outlined in Chapter 2. The use of liquids simply requires a conversion of a known volume of a solution of given concentration into <u>moles</u>. This can always be accomplished by the following equation

$$\text{concentration x volume = amount}$$

or

$$\boxed{\frac{\text{moles}}{\text{liter}} \text{ x liter = moles.}}$$

Once the number of moles of reactants have been determined the problem becomes like those described in Chapter 2 and any of the <u>types</u> may be met. All that is necessary is that two of the three quantities in the above equation be given so the third may be calculated. In the example problems you should always pick out this information first.

A special case of reactions in solution is the <u>titration</u>. A titration is used to determine the amount (moles) of a substance present in a particular form. Let's illustrate with an example. Say

we are investigating acid runoff from a local mine that is killing
vegetation in the surrounding area and we wish to obtain quantitative
data on the acidity of the runoff. The majority of the acid is H_2SO_4
formed by oxidation of sulfides. The acid can be determined according
to the equation $H_2SO_4 + 2NaOH \rightarrow Na_2SO_4 + 2H_2O$. The procedure is as
follows: exactly 25.0 ml of the acidic water was placed in a flask
and a few drops of an indicator added. An indicator is a substance
which signals, usually by a color change, when a reaction is <u>complete</u>.
In this case it signals when all of the acid in the 25.0 ml sample
has been converted to Na_2SO_4 and H_2O. Sodium hydroxide of <u>known</u> concentration is added until the indicator changes color. In this case
exactly 47.4 ml of 0.103 M NaOH was needed to observe the indicator
color change. The apparatus is shown on the next page.

Since we have described the process in detail let's state the
problem again: calculate the molarity of H_2SO_4 in a 25.0 ml sample
if the sample required 47.4 ml of 0.103 M NaOH to titrate to completion. Note the information given: first change the [47.4 ml of
0.103M NaOH] into moles of NaOH and then use the equation for the
reaction. All problems involving solutions require you to change
everything to moles <u>first</u>!

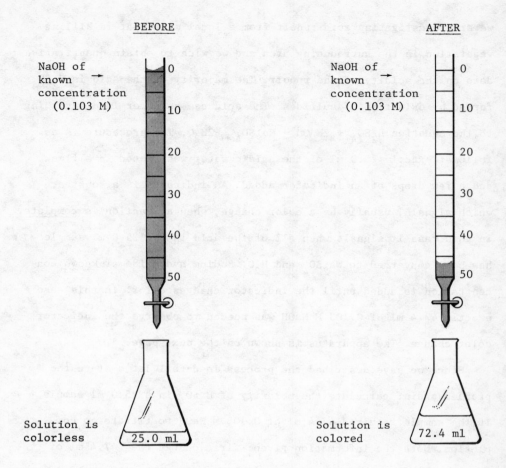

BEFORE

NaOH of known concentration (0.103 M) →

0
10
20
30
40
50

Solution is colorless

25.0 ml

AFTER

NaOH of known concentration (0.103 M) →

0
10
20
30
40
50

Solution is colored

72.4 ml

Now let's calculate the molarity of the acid.

$$H_2SO_4 + 2NaOH \rightarrow Na_2SO_4 + 2H_2O$$

number of moles NaOH used = volume x concentration

$$= 0.0474 \; \ell \times 0.103 \; \frac{moles}{\ell}$$

$$= 4.88 \times 10^{-3} \text{ moles NaOH.}$$

Note the volume has been changed from ml to ℓ by multiplying by 10^{-3} since 1000 ml = 1 ℓ. According to the equation we have 1 mole of H_2SO_4 present for every 2 moles of NaOH used or

$$1 \text{ mole } H_2SO_4 = 2 \text{ mole NaOH}$$

$$\therefore 4.88 \times 10^{-3} \text{ moles NaOH} \times \frac{1 \text{ mole } H_2SO_4}{2 \text{ mole NaOH}} = 2.44 \times 10^{-3} m \ H_2SO_4$$

$\therefore$ we now know that there were 2.44×10^{-3} moles of H_2SO_4

present in our 25.0 ml sample or

$$\frac{2.44 \times 10^{-3} \text{ moles } H_2SO_4}{0.0250 \ \ell} = 0.0976 \text{ M.}$$

Therefore the <u>concentration</u> of the acid, H_2SO_4, in the mine runoff

is 0.0976 M. Other titrations are carried out in the same manner.

We are able in this way to determine the concentration of an unknown

species as long as we know the balanced chemical equation.

For the previous reaction we could have obtained a more detailed

picture in solution by writing the <u>total ionic equation</u>

$$2H^+ + SO_4^{2-} + 2Na^+ + 2OH^- \longrightarrow 2Na^+ + SO_4^{2-} + 2H_2O$$

or the <u>net ionic equation</u>

$$H^+ + OH^- \longrightarrow H_2O.$$

The net ionic equation is obtained by removing all ions that appear

on both sides of the equation i.e. those that do not participate

in the chemical reaction.

B. Types of Problems

1. Problems involving <u>concentrations</u>, <u>moles</u>, and <u>volumes</u>. These
    usually involve the preparation of solutions of known concentra-
    tion.

2. Reactions in solution – solution stoichiometry.

3. Titrations or analysis using solutions.

Examples

1.(i) What is the molarity of a sodium thiosulfate solution prepared
    by dissolving 3.89 g $Na_2S_2O_3$ in water to make 425 ml of
    solution?

$$\text{moles } Na_2S_2O_3 = \frac{3.89 \text{ g}}{158.11 \frac{g}{mole}} = 0.0246 \text{ moles.}$$

∴ we have 0.0246 moles in 425 ml of solution or

$$\frac{0.0246 \text{ moles}}{0.425 \text{ } \ell} = 0.0579 \text{ M.}$$

(ii) Calculate the molarity of commercial nitric acid, HNO_3, which
    is 69% HNO_3 by weight and has a density of 1.42 g ml^{-1}.

    The key to the problem is to convert the density which is

$$[\frac{\text{grams solution}}{\text{ml solution}}] \quad \text{to} \quad [\frac{\text{grams of nitric acid}}{\text{ml of solution}}].$$

    Since we know every 100 g of solution contains 69 g HNO_3,

$$\therefore \frac{1.42 \text{ g solution}}{1 \text{ ml solution}} \times \frac{69 \text{ g}HNO_3}{100 \text{ g solution}} = \frac{0.98 \text{ g } HNO_3}{\text{ml solution}}$$

    Now it is a simple matter to convert g HNO_3 to moles and ml
    solution to liters:

$$\frac{0.98 \text{ g } HNO_3}{\text{ml solution}} \times \frac{1 \text{ mole } HNO_3}{63.01 \text{ g } HNO_3} \times \frac{1000 \text{ solution}}{\text{liter}} = 16 \text{ M.}$$

36

Try the same problem with commercial acetic acid $(C_2H_4O_2)$ which is

100% acetic acid and has a density of 1.05 g ml^{-1}.

(iii) How would you prepare 125 ml of 0.425 M NaCl from 0.675 M NaCl?

Since the <u>final</u> solution is 125 ml of 0.425 M NaCl, we need to

know how many moles of NaCl are contained in the 125 ml. Since

concentration x volume = amount, $0.425 \frac{mole NaCl}{\ell} \times 0.125 \ \ell = 0.0531$

moles NaCl. We can only obtain this number of moles from the

0.675 M NaCl solution. Again, since concentration x volume = amount

$$volume = \frac{amount}{concentration} = \frac{0.0531 \ moles \ NaCl}{0.675 \ \frac{moles \ NaCl}{liter}} = 0.0787 \ liters.$$

Therefore 78.7 ml of 0.675 M NaCl are diluted to 125 ml with water

to give a 0.425 M solution.

2.(i) What is the hydrogen ion concentration of a solution prepared by

mixing 75.0 ml of 0.380 M NaOH with 125 ml of 1.06 M HClO$_4$ and

diluting with water to a final volume of 500 ml? First we need

to realize that NaOH and HClO$_4$ are strong electrolytes forming

Na$^+$, OH$^-$ and H$^+$, ClO$_4$$^-$ ions in solution. Thus the <u>total ionic</u>

<u>equation</u> is

$$Na^+ + OH^- + H^+ + ClO_4^- \longrightarrow Na^+ + ClO_4^- + H_2O$$

The <u>net ionic equation</u> is

$$H^+ + OH^- \longrightarrow H_2O.$$

$$moles \ NaOH = moles \ OH^- = 0.380 \ \frac{mole}{\ell} \times 0.0750 \ \ell = 0.0285$$

$$moles \ HClO_4 = moles \ H^+ = 1.06 \ \frac{mole}{\ell} \times 0.125 \ \ell = 0.133$$

We have then initially 0.113 moles H^+ and 0.0285 moles OH^- and

finally (0.113 - 0.0285) moles H^+ and 0 moles OH^-.

moles H^+ = 0.113 - 0.0285 = 0.105 moles H^+.

$\therefore$ we have 0.105 moles H^+ left in a total final volume of

500 ml or

$$[H^+] = \frac{0.105 \text{ moles}}{0.500 \; \ell} = 0.210 \frac{\text{moles}}{\text{liter}}$$

(ii) How many grams of solid $KClO_4$ will be formed if 200 ml of 0.300 M

KCl are added to 500 ml of 0.400 M $NaClO_4$?

$$KCl + NaClO_4 \xrightarrow{\hspace{1.5cm}} KClO_4 \downarrow + NaCl.$$

Two pieces of information are given therefore we have a limiting

reagent problem involving solutions. Convert each quantity to

moles by using concentration x volume = amount.

$$\text{moles } KCl = 0.300 \frac{\text{mole}}{\ell} \times 0.200 \; \ell = 0.0600 \text{ moles}$$

$$\text{moles } NaClO_4 = 0.400 \frac{\text{mole}}{\ell} \times 0.500 \; \ell = 0.200 \text{ moles}$$

Since 1 mole KCl reacts with 1 mole $NaClO_4$, the limiting reagent is KCl.

$$\therefore 0.0600 \text{ moleKCl} \times \frac{1 \text{ moleKClO}_4}{1 \text{ moleKCl}} = 0.600 \text{ moles } KClO_4 \text{ formed}$$

$$0.0600 \text{ moles } KClO_4 \times \frac{138.55 \text{ gKClO}_4}{\text{moleKClO}_4} = 8.31 \text{ grams } KClO_4.$$

3. Dissolved oxygen, O_2, in water can be determined by a test based on

the following equations:

$$2Mn^{2+} + 4OH^- + O_2 \longrightarrow 2MnO_2 + 2H_2O \qquad (1)$$

$$MnO_2 + 2I^- + 4H^+ \longrightarrow Mn^{2+} + I_2 + 2H_2O \qquad (2)$$

$$I_2 + 2S_2O_3^{2-} \longrightarrow S_4O_6^{2-} + 2I^- \qquad (3)$$

In the first equation manganese(II) ion, Mn^{2+}, is oxidized by the dissolved oxygen in alkaline solution. The resulting solution is acidified and allowed to react with added I^- which is oxidized to I_2 by the MnO_2 (equation (2)). The I_2 can then be determined by titration with sodium thiosulphate, $Na_2S_2O_3$, using starch indicator as in equation (3). Exactly 200 ml of a treated sample was titrated with 0.0250 M $Na_2S_2O_3$. If the volume of $Na_2S_2O_3$ used was 8.32 ml, what is the dissolved oxygen concentration in the sample expressed in mg liter^{-1}?

Here we can calculate the moles of $S_2O_3^{2-}$ used in the titration and work backwards, equation by equation, to obtain the number of moles of O_2 dissolved in the original 200 ml sample.

$$\text{moles } S_2O_3^{2-} = 0.0250 \frac{\text{moles}}{\ell} \times 0.00832\ell = 0.000208 \text{ moles}$$

$$0.000208 \text{ moles } S_2O_3^{2-} \times \boxed{\frac{1 \text{ mole } I_2}{2 \text{ mole } S_2O_3^{2-}}} \times \boxed{\frac{1 \text{ mole } MnO_2}{1 \text{ mole } I_2}} \times \boxed{\frac{1 \text{ mole } O_2}{2 \text{ mole } MnO_2}}$$

$$\qquad\qquad\text{equation (3)} \qquad\qquad \text{equation (2)} \qquad\qquad \text{equation (1)}$$

$$= 0.0000520 = 5.20 \times 10^{-5} \text{ moles } O_2.$$

We only use the connecting links between each equation.

$$\text{mass of } O_2 = 5.20 \times 10^{-5} \text{ moles } O_2 \times \frac{32.0 g O_2}{\text{mole } O_2} = 1.66 \times 10^{-3} g O_2.$$

Therefore, we have $1.66 \times 10^{-3} g O_2$ in 200 ml of solution.

$$\therefore \text{ we have } \frac{1.66 \times 10^{-3} g O_2}{0.200 \ell} = 8.30 \times 10^{-3} \frac{g}{\ell}$$

$$8.30 \times 10^{-3} \frac{g}{\ell} \times \frac{1 \text{ mg}}{10^{-3}g} = 8.30 \frac{\text{mg}}{\ell}.$$

For solutions of water where 1 ml water = $1gH_2O$ the unit $\frac{\text{mg}}{\ell}$ is also parts per million or ppm. Can you work this out?

C.   Test Yourself

Pick out the types of problems and compare with those in chapter 2.

1. (i)   Find the molarity of a solution prepared by diluting 5.0 liters of 6.0 M HCl to 15 liters.

   (ii)   If 4.0 liters of 0.20 M $Ca(NO_3)_2$ and 6.0 liters of 0.80 M $CaCl_2$ are mixed what is the molarity of Ca^{2+}?

   (iii) To what volume must 250 ml of 0.15 M H_2SO_4 be diluted to yield a 0.025 M solution?

   (iv)   Calculate the molarity of commercial hydrochloric acid which is 37% HCl by weight and has a density of 1.19 g ml^{-1}.

   (v)   If 225 ml of a 1.65 M solution of $NaClO_4$ were evaporated to dryness, how many grams of $NaClO_4$ would be left?

2.   Exactly 25.0 g CuS are treated with 140. ml of 2.00 M HNO_3 and the reaction

$$3CuS + 8HNO_3 \longrightarrow 3Cu(NO_3)_2 + 2NO + 4H_2O + 3S$$

occurs.

   (i)   How many grams of $Cu(NO_3)_2$ are produced?

   (ii)   If you had available an unlimited volume of 2.00 M HNO_3 what volume would be required to just react with 25.0 g CuS?

3.   Iron(II) can be determined by titration with dichromate, $Cr_2O_7^{2-}$, according to the equation

$$6Fe^{2+} + 2Cr_2O_7^{2-} + 14H^+ \longrightarrow 6Fe^{3+} + 2Cr^{3+} + 7H_2O.$$

This method was used to determine the percent of iron, added as iron(II) fumarate, in Brand X Multiple Vitamins. One hundred

41

vitamin tablets weighing 38.6 g were dissolved in acid and required 15.2 ml of 0.200 M dichromate to titrate. Calculate the number of mg of iron(II) fumarate, $Fe(C_4H_2O_4)$, in each vitamin tablet and the percent by weight.

4. The pollutant SO_2 can be analysed by passing the SO_2 through a chamber containing O_2 to form SO_3 which is in turn bubbled through water to produce sulphuric acid. The acid can then be titrated with standard NaOH:

$$2SO_2 + O_2 \longrightarrow 2SO_3$$

$$SO_3 + H_2O \longrightarrow H_2SO_4$$

$$H_2SO_4 + 2NaOH \longrightarrow Na_2SO_4 + 2H_2O.$$

A 2.00×10^2 liter sample of air required 14.8 ml of 0.100 M NaOH to titrate the H_2SO_4 produced by the first two equations. Calculate the concentration of SO_2 in the air in g liter^{-1}.

5. A 5.00 g sample of impure $CaCO_3$ was reacted with an excess of 1.000 M HCl (100 ml of 1.000 M HCl) according to the equation $2HCl + CaCO_3 \longrightarrow CaCl_2 + H_2O + CO_2$. After the above reaction was complete, the remaining HCl was titrated with 1.000 M NaOH. The amount of 1.000 M NaOH used was 32.0 ml. Calculate the percent purity of the $CaCO_3$ sample.

6. A solution is prepared by adding 50 ml of 0.184 M NaOH to 50 ml of 0.085 M $HClO_4$ and diluting the resultant mixture with water so that

the final volume of the solution is 500 ml. Calculate the concentrations of: Na^+, ClO_4^-, OH^-, and H^+.

4

Introductory Considerations of Energy and Electromagnetic Radiation

A. Points of Importance

1. The First Law of Thermodynamics

The first law of thermodynamics is a mathematical statement of the principle that energy cannot be created or destroyed i.e. energy is conserved. This is expressed for a system undergoing any process that may involve the transfer of energy as heat or work as

$$\Delta E_{sys} = q + W.$$

For the system and its surroundings

$$\Delta E_{sys} + \Delta E_{surr} = 0.$$

The change in energy of a system from E_1 to E_2, ΔE, is independent of the manner in which the change is accomplished. In other words, any combination

44

of changes in q and w that produce E_2 from E_1 will give the same ΔE.

The most confusing thing about this law as expressed by the equation $\Delta E = q + w$ is the sign (+ or -) of the q and w terms. Remember, + before q or w means <u>heat into</u> and <u>work on</u> the system whereas - before q or w means <u>heat given off</u> and <u>work by</u> the system. These four possibilities are summarized in the table below:

Equation	Type of Activity	Sign of ΔE
$\Delta E = q + w$	energy in as heat into system, energy in as work on system	positive-higher final energy
$\Delta E = - q - w$	energy out as heat given off, energy out as work done by system.	negative, lower final energy
$\Delta E = - q + w$	energy out as heat given off, energy in as work on system	depends on magnitude of q,w
$\Delta E = q - w$	energy in as heat into system, energy out as work done by system	depends on magnitude of q,w

Therefore heat, q, put into the system or work, w, done on the system makes the total energy more <u>positive</u>. And, heat given off by the system or work done by the system makes the total energy more <u>negative</u>. This is shown on the next page.

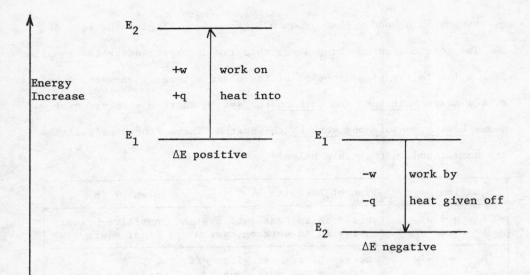

Energy
Increase

E_2 _____

+w   work on

+q   heat into

E_1 _____

ΔE positive

E_1 _____

−w | work by

−q | heat given off

E_2 _____

ΔE negative

    We are all familiar with heat, q, in the above discussion but sometimes are less clear on the work term w. Work is defined as force times the distance through which the force acts. For expansion or compression of a gas, the <u>work done by the gas</u> on the constant pressure surroundings is $\boxed{W = P\Delta V}$ where P is the pressure and ΔV is the volume change. Note that since work is done by the gas PΔV is negative.

2. Enthalpy, H.

    Reactions are normally carried out in vessels open to the atmosphere i.e. at constant pressure. For these reactions the first law can be written (since work is done by the gas):

$$\Delta E = q - w = q - P\Delta V.$$

and $$E_2 - E_1 = q - P(V_2 - V_1)$$

$$q = (E_2 + PV_2) - (E_1 + PV_1)$$

The quantity $E + PV$ can be considered to be the heat absorbed or lost and is given the symbol H i.e. $H = E + PV$. This finally gives

$$q = H_2 - H_1$$

or $$\Delta H = q.$$

H is called the enthalpy and represents the heat content of a state of the system. For a chemical reaction at constant pressure

$$\Delta H = \Sigma H \text{ (products)} - \Sigma H \text{ (reactants)}$$
$$= H \text{ of state 2} - H \text{ of state 1}.$$

If the H of state 2 (products) is greater than the H of state 1 (reactants) then the heat content or "energy content" in the final state is greater than in the initial state and ΔH must be positive. If the H of state 2 is less than the H of state 1 then the heat content in the final state is less than in the initial state and ΔH must be negative. If the sign of ΔH is negative, heat is given off and the reaction is called exothermic. If the sign of ΔH is positive, heat is absorbed and the reaction is called endothermic.

It should be pointed out here that in general individual enthalpies of substances are not available in order to calculate ΔH and another approach is necessary. This involves the defined quantity enthalpy of formation. The enthalpy of formation is the heat evolved or absorbed when one mole of a compound is formed from its elements in their standard

states. The standard state* for a solid or liquid is the pure sub-

stance at one atmosphere external pressure; for a gas it is an ideal

gas at one atmosphere partial pressure (See chapter 7 in the text for

a discussion of partial pressure). The standard state of the reactants

is indicated by the superscript °. i.e. ΔH_f°. Since a great deal of

data has been tabulated at 25°C we arbitrarily think of the elements

and their phases at this temperature. For example, the standard state

of the element H is as the gas H_2 or the standard state of the element

C is as the solid graphite. It is important to specify the state

(solid(s), liquid(ℓ), gas(g)) of the substances involved in the

chemical reaction for this reason. The heat of formation of liquid

ethanol, $C_2H_5OH(\ell)$ can be represented by the equation

$$2C(s) + 3H_2(g) + \tfrac{1}{2}O_2(g) \longrightarrow C_2H_5OH(\ell).$$

By definition, the enthalpy of formation of an element in its standard

state is zero. In the above equation the heat evolved is measured as

66.4 kcal mole^{-1} and we have for this reaction

$$\Delta H^\circ = \Delta H^\circ(\text{products}) - \Delta H^\circ(\text{reactants})$$

$$= \Delta H^\circ_{C_2H_5OH(\ell)} - [2\Delta H^\circ_{C(s)} + 3\Delta H^\circ_{H_2(g)} + \tfrac{1}{2}\Delta H^\circ_{O_2(g)}]$$

$$= -66.4 \text{ kcal mole}^{-1}.$$

---

* For a more detailed discussion of the standard state see

  H. Carmichael, J. Chem. Ed., 53, 695(1976).

Therefore the enthalpy of formation is $\Delta H_f^{\circ}(C_2H_5OH(\ell)) = -66.4$ kcal mole^{-1}. In this way we can obtain a large tabulation of ΔH_f° values which represent the individual enthalpies for use in the above equation. What is the equation representing the enthalpy of formation of $H_2O(\ell)$? of $H_2O(g)$?

Will the magnitude of the numbers obtained be different or the same? Why?

3. <u>Coulomb's Law</u> - the change in potential energy varies inversely with the distance between the two charges.

$$V = \frac{q_1 q_2}{r}$$ where q_1 and q_2 are the two charges, and r is the distance between q_1 and q_2.

4. <u>Electromagnetic Radiation</u>

The properties of light are important as a basis for a discussion of the electron. Electromagnetic radiation consists of photons which have both particle and wave properties. The many familiar forms of electromagnetic radiation, for example, visible light, radio waves, or X-rays, are differentiated by their energy. The energy of a photon is related to the frequency which is in turn related to wavelength and wavenumber by the velocity of light. These relationships are summarized below:

$$E = h\nu$$

$$c = \nu\lambda$$

$$\lambda = \frac{1}{\bar{\nu}}$$

$$\text{and } \therefore E = \frac{hc}{\lambda} = hc\bar{\nu}$$

where c = velocity of light, 3×10^{10} cm sec^{-1}.

ν = frequency in sec^{-1}

h = Planck's constant, 6.626×10^{-27} erg-sec

$\bar{\nu}$ = wave number in cm^{-1}.

These relations can be illustrated schematically by the diagram below for microwave radiation.

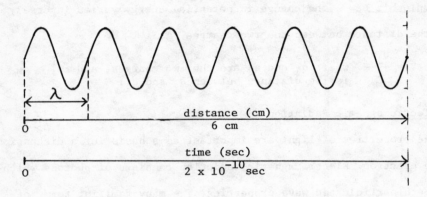

From the diagram we have: λ = 1 cm

$$\nu = \frac{6 \text{ cycles}}{2 \times 10^{-10} \text{sec}} = 3 \times 10^{10} \frac{\text{cycles}}{\text{sec}}.$$

These general considerations are used to explain the line spectrum observed for the emission of radiation from the hydrogen atom. The appearance of radiated light at certain fixed wavelengths means that only certain fixed energies can be emitted from "excited" hydrogen atoms. This leads to the hypothesis that the energies of the excited states are quantized (that is, they can have only certain fixed values), since only certain fixed energies can be radiated from them.

B.   Types of Problems

1.   Problems involving $\Delta E = q + w$ and the sign of all quantities.
     Extension to enthalpy.

2.   Coulomb's Law.

3.   Manipulation of equations relating energy, frequency, and
     wavelength.

Examples

1. (i) Consider a system where 60 calories of heat energy are
       absorbed while the system does 10 calories of work.
       What is the energy change of the system?

$$\Delta E = q - w$$
$$= 60 - 10$$
$$= 50 \text{ calories (Did you get the signs right?)}$$

   (ii) What is the energy change of the system if 35 calories of heat
        energy are transferred by a system to the surroundings while the
        surroundings perform 85 calories of work on the system?

$$\Delta E = -q + w$$
$$= -35 + 85$$
$$= 50 \text{ calories (Did you get the signs correct?)}$$

   Note that in the above examples if both systems are at the same
   energy level initially they can move to the same final energy
   level independently of the path i.e. by different q and w terms.

   (iii) What is the energy change of a system if 50 calories of heat
         energy are added to the system at constant volume?

51

$$\Delta E = q - w$$

$$= 50 - 0$$

$$= 50 \text{ calories.}$$

Here we require a definition of pressure-volume work - the work done by a system as it expands against an applied force (pressure) or the work done on a system as it is compressed by an applied force (pressure). For a constant pressure

$$-w = P\Delta V.$$

$$\text{Since } \Delta V = 0 \quad , -w = 0.$$

(iv) Question 3 at the end of chapter 4.

Calculate the work done when a gas decomposes under a constant pressure of 1.00 atm leading to a volume increase of 150 ml. (1 liter atm = 24.2 cal).

$$-w = P\Delta V$$

$$= 1.00 \text{ atm x } 0.150 \text{ liter}$$

$$= 0.150 \text{ liter atm}$$

But 1 liter atm = 24.2 cal

$$-w = 0.150 \text{ liter atm x } \frac{24.2 \text{ cal}}{1 \quad \text{liter atm}} = 3.63 \text{ cal.}$$

Note that this is a negative w since work is done by the gas.

2. Question 12 at the end of chapter 4.

What is the ratio of the energies of an electron being held by a nucleus if the electron were 1Å away in one case and 3Å away in the other? Which is being held more tightly?

$$V = \frac{q_1 q_2}{r}$$

$$V_1 = \frac{q_1 \times (-1)}{1\overset{\circ}{A}} = -\left(\frac{q_1}{\overset{\circ}{A}}\right)$$

$$V_2 = \frac{q_1 \times (-1)}{3\overset{\circ}{A}} = -\frac{1}{3}\left(\frac{q_1}{\overset{\circ}{A}}\right)$$

$$\frac{V_1}{V_2} = \frac{\left[-\left(\frac{q_1}{\overset{\circ}{A}}\right)\right]}{\left[-\frac{1}{3}\left(\frac{q_1}{\overset{\circ}{A}}\right)\right]} = 3.$$

Since V_1 is more negative than V_2, it is the more stable system and the electron is more tightly held for this case (1A away).

3. One of the strong lines in the atomic spectrum of barium corresponds to emission of photons of energy 4.37×10^{-12} erg. Calculate the wavelength of the emission.

$$\Delta E = \frac{hc}{\lambda}$$

$$\therefore \lambda = \frac{hc}{\Delta E} = \frac{6.63 \times 10^{-27} \text{erg sec} \times 3.00 \times 10^{10} \text{cm sec}^{-1}}{4.37 \times 10^{-12} \text{erg}}$$

$$= 4.55 \times 10^{-5} \text{ cm.}$$

$$\text{or} = 4.55 \times 10^3 \overset{\circ}{A} \text{ (Can you get this?)}$$

C. Test Yourself

1. The following experiment is carried out: a thin-walled aluminum tin
    is half-filled with water and it is heated just enough to make the
    water boil and the tin is sealed off. The tin and its contents are
    allowed to sit (cool) when suddenly the tin collapses.
    In terms of the first law of thermodynamics indicate

    (i) the energy of the system when it is heated to make the water
         boil and

    (ii) explain the collapse of the tin.

2. Write the equation for the heat of formation of the following:

    (i) $CH_3OH(\ell)$

    (ii) $NO(g)$

    (iii) $H_2SO_4(\ell)$

    (iv) $C(diamond)$

    (v) $CaCO_3(s)$

3. Give examples of the following using "chemical" equations:

    (i)   enthalpy of vapourization.

    (ii) enthalpy of fusion.

    (iii) enthalpy of sublimation.

    (iv) enthalpy of combustion.

    (v)   enthalpy of formation.

4. One of the lines in the atomic spectrum of sodium appears at 3303 Å.

    (i) Calculate the energy in ergs and kcal of a photon having this
         wavelength.

54

(ii) Calculate the frequency of the radiation.

5. The energy required to remove an electron from a photoelectric material is 7.4×10^{-12} erg. Calculate the lowest frequency light that can give rise to the photoelectric effect with this material.

6. Light of wavelength 26.9 Å is required to cause electrons to be emitted from a photoelectric material. What is the energy required for electrons to be emitted?
    ($1 \text{ Å} = 10^{-8}$ cm).

5

Atomic Structure, Atomic Properties, and Molecular Consequences of Atomic Properties

A. Concepts of Importance

1. Atomic Structure

The development of the concepts of atomic structure and their equations are briefly summarized below:

(i) Bohr Model: See section 5-1 in the text.

(ii) De Broglie and the Dual Properties of Matter. See section 5-2 in the text.

Particles have both wave and particle properties. The wavelength of a particle is given by

$$\lambda = \frac{h}{mv} \qquad (\text{from } p = mv = \frac{h}{\lambda})$$

where v = speed of particle

p = momentum of particle = mv

m = mass of particle

h = Planck's constant.

(iii) <u>Uncertainty Principle</u>: See section 5-3 in the text.

(iv) Quantum Mechanics: See section 5-4 in the text.

(v) <u>Quantum Numbers</u>

The behaviour of electrons can be described by mathematical equations that have many solutions. These equations correctly predict the experimental properties of electrons and are thus justified. These equations tell us that matter can be represented by a wavefunction ψ. The square of this wavefunction, ψ^2, is related to "probability" and thus provides a physical picture of the atom. This "probability" is a consequence of the uncertainty principle.

Quantum numbers, which you have all heard of, result from the solutions to the mathematical equations. These quantum numbers specify the characteristics of orbitals in atoms and a summary of this information is given in the table on the next page. Compare the information in the table with the actual quantum numbers given in Tables 5-2 of the text.

Quantum Number	Information	Values	Comments
n	related to the average distance of a group of orbitals from the nucleus	1,2,3...	number of electrons in this major group of orbitals = $2n^2$
ℓ	related to the shape of the orbital.	0 to n-1	designates the type of orbitals within the major group $\ell = 0 \rightarrow$ s orbital $\ell = 1 \rightarrow$ p orbital $\ell = 2 \rightarrow$ d orbital $\ell = 3 \rightarrow$ f orbital
m_ℓ	related to the orientation of each orbital in space (the s,p,d etc.)	$-\ell, -\ell+1, \ldots, 0, \ldots, \ell-1, \ell$	the number of values designate the number of each type of orbital (eg) for $\ell = 2$, $m_\ell = 2,1,0,-1,-2$ i.e. 5 values and $\therefore$ 5 orbitals oriented differently in space.

(v) <u>Electronic Configurations</u>

Using the concepts developed for the hydrogen atom (everything

that has been mentioned to this point) it is possible to extend the

model to many-electron atoms. In adding electrons to an atom the

lowest energy orbitals are filled first (the Aufbau order) with

each orbital accommodating two electrons. Most electronic con-

figurations can be written with recourse to the following mimic

scheme:

(1) Write out the orbitals by grouping them in terms of the

quantum number n.

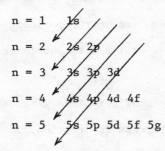

$$n = 1 \quad 1s$$
$$n = 2 \quad 2s \ 2p$$
$$n = 3 \quad 3s \ 3p \ 3d$$
$$n = 4 \quad 4s \ 4p \ 4d \ 4f$$
$$n = 5 \quad 5s \ 5p \ 5d \ 5f \ 5g$$

(2) Draw diagonal lines as shown above and add electrons along the

arrows from top to bottom.

(eg.) Write the electronic configuration of $_{19}K$

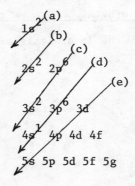

We add 19 electrons starting

at diagonal (a) in the direction

of the arrow. Two electrons are

added to each orbital.

∴ we have

$1s^2 \ 2s^2 \ 2p^6 \ 3s^2 \ 3p^6 \ 4s^1$.

59

The quantum number required to distinguish the two electrons in each orbital is the spin quantum number, m_s, which can have the value $+ \frac{1}{2}$ or $- \frac{1}{2}$. The two electrons that go into any one orbital must have opposite spin quantum numbers. With the introduction of the <u>Pauli exclusion principle</u> - no two electrons in an atom can have all four quantum numbers alike - we can now uniquely label all electrons in an atom. It should be remembered, however, that the first three quantum numbers actually refer to the orbital that contains the electron. The table on the following page relates the quantum numbers to the atoms. Spend some time relating this table of quantum numbers to the <u>periodic table</u> and to the order of filling up of orbitals to obtain the electronic configuration.

In addition to the electronic configurations given above we can use a box diagram to indicate this information. This method of writing electron configurations brings out a very important rule, called Hund's rule. <u>Hund's rule states that when electrons are added to degenerate orbitals</u> (eg. to the three 2p orbitals) <u>one electron is placed in each degenerate orbital before pairing electrons</u>. For example the box diagram for the 2p orbitals of N is

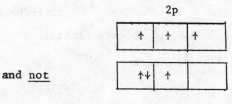

and <u>not</u>   .

60

Table of Quantum Numbers

n	ℓ	m_ℓ	m_s		group of elements
1	0	0	$\pm\frac{1}{2}$		H, He
2	0	0	$\pm\frac{1}{2}$		Li, Be
	1	1	$\pm\frac{1}{2}$	)	
		0	$\pm\frac{1}{2}$	}	B, C, N, O, F, Ne
		-1	$\pm\frac{1}{2}$	)	
3	0	0	$\pm\frac{1}{2}$		Na, Mg
	1	1	$\pm\frac{1}{2}$	)	
		0	$\pm\frac{1}{2}$	}	Al, Si, P, S, Cl, Ar
		-1	$\pm\frac{1}{2}$	)	
	2	2	$\pm\frac{1}{2}$	)	
		1	$\pm\frac{1}{2}$	)	
		0	$\pm\frac{1}{2}$	}	Sc, Ti, V, Cr, Mn, Fe, Co, Ni, Cu, Zn
		-1	$\pm\frac{1}{2}$	)	
		-2	$\pm\frac{1}{2}$	)	

2.   Periodic Properties.

When chemical reactions take place, only the outermost electrons generally participate. For this reason the chemical properties of an element are largely associated with the electron configuration of the outermost shell. The electrons in the outermost shell are called the valence electrons. The periodic table is arranged in groups of elements with similar outer electronic configurations as illustrated on the periodic

61

table on page 63. On the second periodic table the group numbers and the common names of some groups of families are given.

The arrangement of the periodic table as given above allows us to discuss the variation of some important atomic properties such as atomic radii, ionisation energy, electron affinity, and electronegativity.

Atomic Radii – Trends I (page 63).

(a) Atomic radii <u>decrease</u> on going from <u>left to right</u> on the periodic table. This is due to the fact that the same principal shell is involved but the nuclear charge increases and thus the electrons are pulled closer to the nucleus.

(b) Atomic radii increase on going from <u>top to bottom</u> on the periodic table. This is due to an increase in principal shell, n, as one moves down a group.

(c) <u>Size and Charge</u>

 (i) If an electron is removed from a neutral atom the nuclear charge attracts a lesser number of electrons and pulls them closer to the nucleus. Therefore, for example, $K > K^+$.

 (ii) If an electron is added to a neutral atom the nuclear charge attracts a greater number of electrons with less facility and expansion results. Therefore, for example, $F^- > F$.

62

Electronic Configuration and the Periodic Table

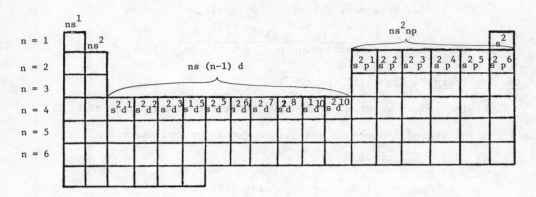

n = 1 ns^1 ns^2

$ns(n-1)d$

ns^2np

s^2

n = 2 s^2p^1 s^2p^2 s^2p^3 s^2p^4 s^2p^5 s^2p^6

n = 3

n = 4 s^2d^1 s^2d^2 s^2d^3 s^1d^5 s^2d^5 s^2d^6 s^2d^7 s^2d^8 s^1d^{10} s^2d^{10}

n = 5

n = 6

Group Numbers and Common Names

1A–ALKALI METALS

11A–ALKALINE EARTHS

MAIN GROUP

HALOGENS

V111A–NOBLE GASES

111A   1Va   VA   V1A   V11A

TRANSITION

Periodic Trends I – Atomic Radii

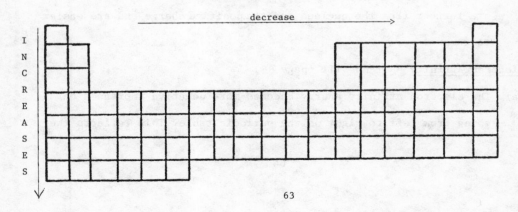

decrease →

INCREASES

63

(iii)   For an <u>isoelectronic</u> series (same number of electrons)

the radius varies directly as the charge on the atom.

$$O^{2-} > F^- > Ne > Na^+ > Mg^{2+} > Al^{3+} > P^{5+}$$
$$\overrightarrow{\text{decreasing size, increasing positive charge}}$$

<u>Ionisation Energy</u> - Trends II (page 65)

Each electron in an atom has its own ionisation energy. For example the first and second ionisation energies (removal of <u>one</u> and <u>two</u> electrons) of magnesium are represented by

$$Mg \longrightarrow Mg^+ + e^- \quad \text{1st I.E.}$$
$$Mg^+ \longrightarrow Mg^{2+} + e^- \quad \text{2nd I.E.}$$

The second ionisation energy is greater than the first since it is more difficult to separate an electron from a positively charged species than from a neutral species. The following trends pertain to the <u>first I.E.</u>

(a)   The ionisation energy increases on going from left to right on the periodic table. This is due to the fact that the same principal shell is involved but the nuclear charge increases thus making it more difficult to remove an electron.

(b)   The ionisation energy decreases on going from top to bottom on the periodic table. The principal shell, n, increases and the electrons are further from the nucleus and the positive charge and are easier to remove.

<u>Electron Affinity</u> - Trends III (page 65)

(a)   The electron affinity energy becomes more negative increases on going from left to right on the periodic table. This reflects the

Periodic Trends II – Ionisation Energy

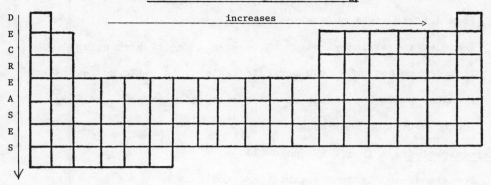

Periodic Trends III – Electron Affinity

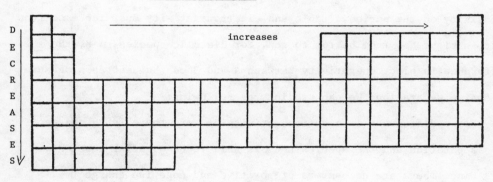

Periodic Trends IV – Electronegativity

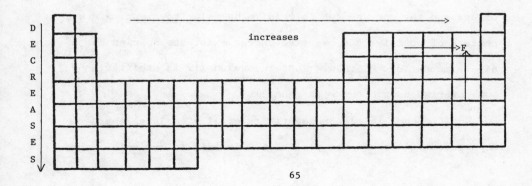

the decreasing radii and greater electron attraction as one goes from left to right on the table.

(b)   The electron affinity energy becomes less negative on going from top to bottom on the periodic table. This reflects the increasing radii and lesser electron attraction as one goes from top to bottom on the table.

Electronegativity - Trends IV (page 65).

The trends in electronegativity are the same as those for electron affinity. They are easy to remember since F has the highest electronegativity on the periodic table and electronegativity must increase from left to right and from bottom to top. For diatomic species, A–B, the difference in electronegativity between A and B can be used as a measure of charge separation. We can predict the following:

1)   For a difference in electronegativity of zero there is no charge separation and the species A–B is said to be nonpolar covalent. Since there are no centers of positive and negative charge the molecule has no dipole moment. eg. Cl_2.

2)   For a difference in electronegativity of <1.9, charge separation exists and the species A–B is said to be polar covalent. Since centers of positive and negative charge exist the species has a dipole moment. eg. HCl. Note that the value 1.9 is arbitrary and some overlap does occur with case 3).

3)   For a difference in electronegativity of >1.9, a large charge separation exists and the species A–B is said to be ionic.

66

Usually the charge separation is regarded as complete (a full +
and − on each species) in this case e.g. K^+Cl^-. One can readily
appreciate from the periodic trends in electronegativity that
case 3) would be for atoms widely separated on the table while
case 2) would be for atoms reasonably close together on the table.
Again some overlap occurs with case 2).

When we consider molecules larger than diatomics we need to
know the geometry before we can predict whether the molecule will
have a dipole moment. We will consider this in chapter 6.

3. Lewis Structures

Lewis structures are extremely important because they give a
first approximation to the electronic structures of many molecules and
ions and they allow us to predict molecular geometries. Rather than
restate the rules for writing Lewis structures we will do a few
of the more complicated examples in detail to illustrate the rules.

(i) Draw the Lewis structure of NO^+.

(1) Draw the skeleton structure: $\longrightarrow$ $\boxed{N-O^+}$

(Remember the single line represents 2 electrons)

(2) Count the total number of valence electrons, V:

V = 5 (from nitrogen) + 6 (from oxygen) − 1 (positive charge)

= 10.

(Try to remember the valences from the group numbers on the
periodic table).

67

(3)   Subtract 2 electrons from V for each line drawn in the skeleton
structure and distribute the remaining electrons so as to place
eight around each atom if possible:

10 (valence electrons) - 2 (1 line) = 8 left

$[: \ddot{\underset{..}{N}} - \ddot{O}^+]$ You will find it is not possible (using

the 8 electrons left) to place 8 electrons

around each atom as shown at the left.

(4)   We need to use multiple bonds.* Make a double bond (2 lines)
and repeat step (3).

N=O $^+$

10 (valence electrons) - 4 (2 lines) = 6 left.

* When using multiple bonds in writing Lewis structures always
make one double bond first, then two double bonds if necessary,
or a triple bond.

Again as shown [$\overset{..}{N} = \overset{..}{\underset{..}{O}} ^+$] it is not possible (using the
6 electrons left) to place 8 electrons around each atom.

(5)   Make a triple bond (3 lines and repeat step (3)).

N $\equiv$ O $^+$

10 (valence electrons) -6 (3 lines) = 4 left

∴ $[: N \equiv O :^+]$. In this case it is possible to place 8

electrons around each atom. When counting the electrons the
"lines" represent shared electrons and are counted for each
atom i.e. around N we have two dots (2 electrons) and three

68

lines (6 electrons) for a total of eight. Similarly, around O
we have three lines (6 electrons) and two dots (2 electrons)
for a total of eight.

For the next examples let's simply tabulate our results as we go
along and leave out the words.

(ii) Draw the Lewis structure for NO_2^+.

$[O - N - O]^+$ $V = 6 + 5 + 6 - 1 = 16$ electrons

$[:\ddot{O} - \ddot{N} - \ddot{O}]^+$ $(16 - 6) = 12$ left $\rightarrow$ NO.

$[:\ddot{O} - \ddot{N} = \ddot{O}]^+$ $(16-6) = 10$ left $\rightarrow$ NO.

$\left[\ddot{O} = N = \ddot{O}\right]^+$ $(16 - 8) = 8$ left $\rightarrow$ YES!

You should be going through the "words" as in example (i) for
the above in your head.

(iii) Draw the Lewis structure for PO_4^{3-}.

$$\left[:\ddot{O}: \atop :O - P - O: \atop :O: \right]^{3-}$$

$V = 6 + 6 + 6 + 6 + 5 + 3 = 32$ e$^-$'s

$(32 - 8) = 24$ left $\rightarrow$ YES!

(iv) This example illustrates the concept of <u>resonance</u>.
Draw the Lewis structure for CO_3^{2-}.

$$\left[\begin{matrix} O \\ C \\ O \quad O \end{matrix} \right]^{2-}$$

$V = 6 + 6 + 6 + 4 + 2 = 24$ electrons

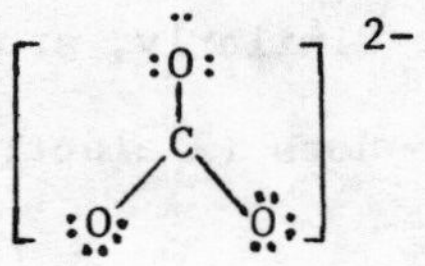

$$\left[\begin{array}{c} \overset{\cdot\cdot}{\underset{\cdot\cdot}{O}} \\ C \\ :\overset{\cdot}{O}: \quad :\overset{\cdot\cdot}{O}: \end{array} \right]^{2-} \qquad (24 - 6) = 18 \text{ left} \rightarrow \text{NO}$$

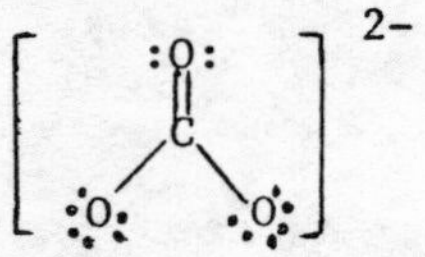

$$\left[\begin{array}{c} :\overset{\cdot\cdot}{O}: \\ C \\ :\overset{\cdot}{O}: \quad :\overset{\cdot}{O}: \end{array} \right]^{2-} \qquad (24 - 8) = 16 \text{ left} \rightarrow \text{YES!}$$

However two other possibilities also exist,

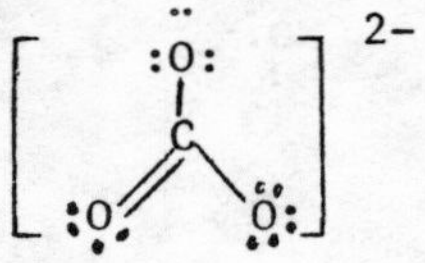

$$\left[\begin{array}{c} :\overset{\cdot\cdot}{O}: \\ C \\ :\overset{\cdot}{O}: \quad :\overset{\cdot\cdot}{O}: \end{array} \right]^{2-} \quad \text{and} \quad \left[\begin{array}{c} :\overset{\cdot\cdot}{O}: \\ C \\ :\overset{\cdot\cdot}{O}: \quad :\overset{\cdot}{O}: \end{array} \right]^{2-}$$

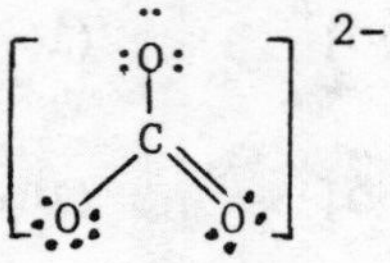

since there is no reason why only one particular oxygen of the three
identical oxygens should form a double bond. Experiment shows that
none of the individual structures above is correct but that all
carbon-oxygen bonds are the same. The real structure is a resonance
hybrid of all three of the structures above and can be represented as

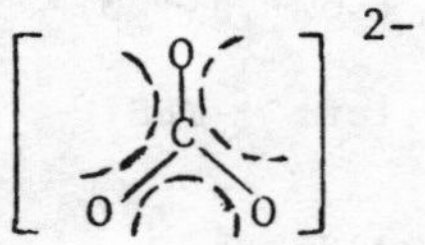

$$\left[\begin{array}{c} O \\ C \\ O \quad O \end{array} \right]^{2-}$$

where the dashed lines show that the C-O
bonds are all the same (something between
a double and single bond).

The bond order is one for a single bond, two for a double bond
and three for a triple bond. Therefore in the above example the
C-O bond order can be said to be 1.33 since the actual bond   is

70

from 1 double bond distributed among 3 C-O's. The following
table gives some information on the main types of bonds.

General Type of Bond	Pairs of Electrons	Name of Bond(s) Making up General Type	Example
single	1	sigma, σ	H_2, CH_4
double	2	1 sigma plus 1 pi, π	O_2, C_2H_4
triple	3	1 sigma plus 2 pi	N_2, C_2H_2

B.   Types of Problems.

1.   Atomic structure, quantum numbers, and electron configurations.

2.   Periodic properties.

3.   Lewis structures.

Examples

1.(i) Question 11. from the end of chapter 5.

List the elements that have box diagrams similar to nitrogen.

The box diagram for nitrogen is

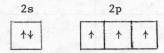

The elements that have similar diagrams are the group VA elements
i.e. all those elements below N: P, As, Sb, Bi. Look up the electron
configurations to verify this.

(ii) Question 13. from the end of chapter 5.

How many solutions to the Schrödinger equation for a hydrogen atom
are there with energies corresponding to n = 2? to n = 4?

The number of solutions is given by n^2 and corresponds to the number
of orbitals for a particular n value.

n = 2 , n^2 = 4 $\therefore$ 4 solutions and 4 orbitals with n = 2 i.e. 2s,

$2p_x$ $2p_y$ $2p_z$.

n = 4 , n^2 = 16 $\therefore$ 16 solutions and 16 orbitals with n = 4 i.e. 4s,

$4p_x$ $4p_y$ $4p_z$, five 4d orbitals, and seven 4f orbitals.

(iii) Which of the following sets of quantum numbers represents an impossible arrangement?

72

(a) $n = 2$, $\ell = 0$, $m_\ell = 0$, $m_s = +\frac{1}{2}$

(b) $n = 5$, $\ell = 4$, $m_\ell = -1$, $m_s = +\frac{1}{2}$

(c) $n = 3$, $\ell = 1$, $m_\ell = +1$, $m_s = +\frac{1}{2}$

(d) $n = 3$, $\ell = 3$, $m_\ell = -1$, $m_s = -\frac{1}{2}$

Check each set of quantum numbers:

(a) $n = 2$ ∴ $\ell = 1, 0$ for $\ell = 0$, $m_\ell = 0$, $m_s = \pm\frac{1}{2}$ O.K.

(b) $n = 5$ ∴ $\ell = 4$, $3,2,1,0$ for $\ell = 4$, $m_\ell = 4,3,2,1,0, -1$, $-2,-3,-4$,

$m_s = \pm\frac{1}{2}$ O.K.

(c) $n = 3$ ∴ $\ell = 2, 1$, 0 for $\ell = 1$, $m_\ell = 1$, 0, -1, $m_s = \pm\frac{1}{2}$ O.K.

(d) $n = 3$ ∴ $\ell = 2, 1, 0$ <u>not $\ell = 3$</u> ∴ this set is not possible.

(iv) Write the electronic configuration for $_{34}Se^{2-}$. Since it is Se^{2-} there

are a total of $34 + 2 = 36$ electrons to distribute. This is then the

electronic configuration for Kr. (atomic number 36 on periodic table).

Simply add 36 electrons to the mimic scheme:

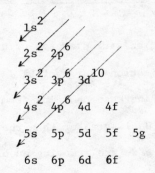

∴ we have $1s^2\ 2s^2\ 2p^6\ 3s^2\ 3p^6\ 4s^2\ 3d^{10}\ 4p^6$.

(v) Write the electronic configuration for $_{24}Cr$.

Application of the mimic scheme gives

$1s^2\ 2s^2\ 2p^6\ 3s^2\ 3p^6\ 4s^2\ 3d^4$ but this is <u>not correct</u>.

73

It turns out that this is an exception and the more stable electronic configuration is $1s^2\ 2s^2\ 2p^6\ 3s^2\ 3p^6\ 4s^1\ 3d^5$.

This simply reflects the fact that the 4s and 3d orbitals are very close in energy for Cr and the <u>half-filled</u> orbitals are very stable. In the first transition series there is one more exception, the electronic configuration for $_{29}$Cu. Look it up.

2(i) The following graph shows a plot of first ionisation energy versus atomic number.

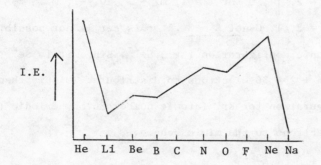

(a) Explain why the first ionisation energy of He is greater than that of Ne and why the ionization energy of Li is greater than that of Na.

(b) Explain the irregular increase on going from Li to Ne.

     (a) The I.E. of He is greater than that of Ne because in He we are removing an electron from n = 1 whereas in Ne the electron is less tightly held, n = 2. The I.E. of Li is greater than that of Na for the same reason: Li (n = 2) and Na (n = 3).

(b) The overall increase of ionisation energy as we go from
Li to Ne is due to the increasing nuclear charge in the
same principal shell n = 2. An explanation for the drop
at Be → B and at N → O can be found in the electronic
configurations of the elements.

	2S	2P		
Be	↑↓			
B	↑↓	↑		
N	↑↓	↑	↑	↑
O	↑↓	↑↓	↑	↑

B has a slightly smaller ionisation energy than Be because the
electron being removed is a 2p electron which is of slightly higher
energy than the paired 2s electron in Be. The text discusses the
reasons for this small difference in energy in terms of the ψ^2
plots.

In O one of the 2p orbitals contains a paired set of electrons
and electron repulsions make this a higher energy state than if
they were not paired as in N. Thus the electron is easier to
remove in O.

(ii) Which of the species below has the smallest radius?

(a) Ar   (b) Br^- (c) Ca^{2+} (d) Mn^{7+} (e) S^{2-}.

The first thing to realize here is that the series Mn^{7+}, Ca^{2+},
Ar, S^{2-} is an <u>isoelectronic</u> series. Thus in this group Mn^{7+} would
have the smallest radius since it has the highest nuclear charge.

Now what about Br^-? This is easily disposed of since Br^- is larger than Cl^- which would be a member of the isoelectronic series above. This simply reflects the fact that Br^- has a partially filled $n = 4$ shell whereas all the other species have a filled $n = 3$ shell which is closer to the nucleus than the $n = 4$ shell. Therefore the correct answer is Mn^{7+}.

(iii) Question 23. and 24. at the end of chapter five require you to use the periodic trends and the positions of the elements in the table. We will simply state the correct answers.

Question 23. Indicate the more electronegative element of the following pairs:

(a) C or Si?   C

(b) B or C?   C

(c) Si or Pb?   Si

(d) V or Cu?   Cu

Question 24. Indicate the larger element (or ion) in the following pairs:

(a) S or S^{2-}? S^{2-}

(b) Al or C?   Al

(c) Al^{1+} or Al^{3+}? Al^{1+}

(d) Mn^{2+} or Cu^{2+}? Mn^{2+}

(e) Si or Pb?   Pb

(f) Na or Mg?   Na.

3. Since we have already covered a number of examples in A. we
    will only do two more.

(i) Draw the Lewis structure for NCS⁻.

$$[N\text{-}C\text{-}S]^-$$ $V = 5 + 4 + 6 + 1 = $ 16 electrons.

$$[:\overset{..}{\underset{..}{N}}\text{-}\overset{..}{C}\text{-}\overset{..}{S}]^-$$ $(16-4) = 12$ left → NO.

$$[.\overset{..}{\underset{.}{N}}=\overset{..}{C}\text{-}\overset{..}{S}:]^-$$ $(16-6) = 10$ left → NO.

$$[:\overset{.}{N}=C=\overset{.}{\underset{..}{S}}:]^-$$ $(16-8) = 8$ left → YES.

We also can write 2 more structures with triple bonds:

$$[:N \equiv C - \overset{..}{\underset{..}{S}}:]^-$$ and $$[:\overset{..}{N} - C \equiv S:]^-$$

The three structures are resonance forms which contribute
unequally to the correct structure.

(ii) Draw the Lewis structure for XeF_2.

     F – Xe – F $V = 7 + 8 + 7 = 22$ electrons

$$:\overset{..}{\underset{..}{F}} - \overset{..}{\underset{.\,.}{Xe}} - \overset{..}{\underset{..}{F}}:$$ $(22 - 4) = 18$ left

Note that Xe has 10 electrons around it and it is not possible
to attain an octet.

77

C. Test Yourself

1. Circle the appropriate answer.

   (i)   Unpaired electrons are present in ground-state atoms of

         (a) Ne   (b) Mg   (c) Be   (d) N   (e) Zn

   (ii) Which of the following is/are isoelectronic with K^+?

         (a) Cl^- (b) Kr   (c) Ne   (d) Mg^{2+} (e) F^-.

   (iii) The number of m_ℓ values possible for a 3d electron is

         (a) 1   (b) 3   (c) 4   (d) 5   (e) 7.

   (iv) Which atom contains an electron with quantum numbers n = 3,

         ℓ = 2, m_ℓ = 0, m_s = ½?

         (a) K   (b) Mg   (c) Cl   (d) Ne   (e) Co.

   (v)   A possible set of quantum numbers for the 4s electron in K is

         (a) n = 3, ℓ = 2, m_ℓ = 0, m_s = ½.

         (b) n = 4, ℓ = 0, m_ℓ = 0, m_s = ½.

         (c) n = 4, ℓ = 1, m_ℓ = 0, m_s = ½.

         (d) n = 4, ℓ = 2, m_ℓ = 0, m_s = ½.

         (e) n = 4, ℓ = 3, m_ℓ = 0, m_s = ½.

   (vi) The atom that has the ground-state electronic configuration

         $3s^2 3p^3$ for the outer shell is

         (a) As   (b) N   (c) Si   (d) P   (e) Ne.

   (vii) Which of the following elements has the highest first

         ionisation energy?

         (a) Li   (b) Be   (c) Na   (d) K   (e) Rb

78

(viii) The p atomic orbitals are oriented in space with respect

to each other at angles of

(a) 45°   (b) 90°   (c) 109°   (d) 120°   (e) 180°.

(ix) The element with the most metallic character is

(a) Al   (b) K   (c) Sr   (d) Cs   (e) Fe.

(x) Which of the following compounds contains bonds with the most

ionic character?

(a) AsF_5 (b) GaF_3 (c) CaF_2 (d) GeF_4 (e) KF

2.   Write the electronic configuration and the box diagram for

(i) $_{28}Ni$

(ii) $_{24}Cr^{3+}$

(iii) $_{28}Ni^{2+}$

(iv) $_{20}Ca^{2+}$

(v) $_{35}Br$

3.   Draw Lewis structures for the following and include all important

resonance forms. The central atom is underlined.

(i) $F_2\underline{O}$

(ii) $\underline{N}H_4^+$

(iii) $H\underline{N}O_3$ (O–N⟨$^O_{OH}$)

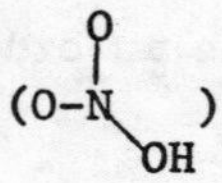

(iv) $N\underline{C}O^-$

(v) $\underline{S}O_3^{2-}$

(vi) N_3^-

(vii) $\underline{N}O_2^-$

(viii) $\underline{P}Cl_3$

(ix) $\underline{C}S_2$

(x) C_2H_4 skeleton is H_2C-CH_2

4. (i) Name two ions having each of the following electronic configurations:

   (a) $1s^2\ 2s^2\ 2p^6\ 3s^2\ 3p^6$

   (b) $1s^2\ 2s^2\ 2p^6\ 3s^2\ 3p^6\ 3d^{10}$

   (c) $1s^2$

(ii) Which of the following electron configurations correspond to atoms in their <u>ground state</u>, which are <u>excited state</u> electron configurations, and which are <u>impossible</u>?

$$1s^2\ 2d^1\ 2p^1$$
$$1s^2\ 2s^2$$
$$1s^2\ 2s^3$$
$$1s^2\ 2s^2\ 2p^2$$
$$1s^2\ 2s^2\ 2p^6\ 2d^1$$
$$1s^2\ 2s^2\ 2p^3$$
$$1s^2\ 2s^1\ 3s^1$$

5. Arrange the following groups of atoms and ions in order of decreasing radius:

   (a) C, Si, Ge.

   (b) B, Li, Be.

   (c) S^{2-}, Ca^{2+}, Cl^-, K^+.

6

Molecular Structure

A. Points of Importance

1. Ionic Bonding: See section 6-1 in the text.

2. Lewis Structures, Geometries, and Hybridisation.

The geometry of molecules and ions can be predicted using Lewis
structures. The basic idea is that the geometry of a molecule is
determined by the number of electron pairs around the central atom.
The geometries depend on whether the central atom uses all of the
valence electrons in forming sigma bonds or whether the central atom
contains lone pairs or nonbonding pairs of electrons. In the table on
pages 83 and 84 the areas in the rectangles represent the cases where
all the electron pairs around the central atom are bonding electron
pairs. (BP). This is rule 1 and 4. in the text. For the remaining
entries the central atom also contains lone pairs of electrons (LP)

or <u>nonbonding pairs</u> of electrons. The latter situation is covered by rule 2 and 4 in the text. <u>Remember</u> you cannot "see" the <u>lone pairs</u> thus they are not used in describing the actual shape of the molecule.

This approach to predicting the shape only considers the sigma orbitals but can be used for molecules which contain double or triple bonds. The π electrons are simply not counted as groups or lone pairs. A few examples will illustrate this point.

(1) Predict the shape of SO_2.

First write the Lewis structure. The two resonance structures are:

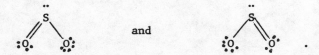

Since we only consider sigma bonds we have 2BP and 1LP around the central sulfur atom. Therefore we have

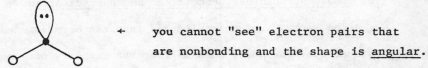

← you cannot "see" electron pairs that are nonbonding and the shape is <u>angular</u>.

(2) Predict the shape of C_2H_2.

The Lewis structure is   H—C≡C—H.

Thus around each carbon we have   2BP <u>i.e.</u>

H ⌐ C ╀ C and C ╀ C ╀ H

For 2BP the shape is <u>linear</u>.

From a knowledge of the geometry of a molecule we can predict whether the molecule has a dipole moment (whether the molecule is polar). In

Total Number of Electron Pairs Around Central Atom	Orbital Shape	Orbital Shape and Lewis Structure	Actual Shape	Example
2	linear	2BP	linear	BeF_2
3	trigonal planar	3BP	trigonal planar	BF_3
3		2BP, 1LP	angular	$SnCl_2$
4	tetrahedral	4BP	tetrahedral	CCl_4
4		3BP, 1LP	trigonal pyramidal	NH_3
4		2BP, 2LP	angular	H_2O
5	trigonal bipyramidal	5BP	trigonal bipyramidal	PCl_5

83

Total Number of Electron Pairs Around Central Atom	Orbital Shape	Orbital Shape and Lewis Structure	Actual Shape	Example
5	trigonal bipyramidal	4BP, 1LP	Sawhorse	SF_4
5		3BP, 2LP	Bent T	BrF_3
5		2BP, 3LP	linear	ICl_2^-
6	octahedral	6BP	octahedral	SF_6
6		5BP, 1LP	square pyramidal	IF_5
6		4BP, 2LP	square planar	XeF_4

84

molecules that don't contain lone pairs of electrons and all the groups attached to the central atom are the same, the polarities of the bonds will exactly cancel out and there is no dipole moment. i.e. CCl_4 which is tetrahedral has no dipole moment. Another way of stating this is to say that the molecule is exactly symmetrical around the central atom. If the symmetry is destroyed as in $CHCl_3$ (again tetrahedral) the polarities of the individual bonds do not cancel and the molecule has a dipole moment. All molecules that are not symmetrical (for example H_2O and SO_2 are both angular) have dipole moments. The same principles apply to other geometries.

Up to this point we have conveniently avoided a discussion as to which atomic orbitals around the central atom are involved in bonding to the other groups. Sometimes predictions of the <u>type</u> of compound and the <u>shape</u> of the molecule fail when the atomic orbitals are used. For example let's predict the type of compound expected from C and H. Since the electronic structure of carbon is $1s^2\ 2s^2\ 2p^2$ and the box structure for the valence electrons is

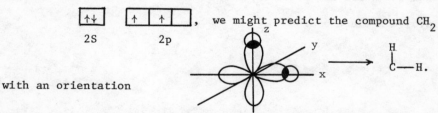

, we might predict the compound CH_2

with an orientation

This is of course incorrect since we know the correct formula is CH_4 which has a tetrahedral shape. How can we rationalize these facts? We do so after the fact with the justification that it works. Once the shape or geometry of a molecule is known a mathematical procedure called <u>hybridisation</u> can be applied to give <u>new atomic orbitals</u> on the central atom with the correct orientation to overlap and form bonds. This procedure is discussed in the text in more detail. The table on the next page shows the relation between the "geometry of a molecule" and the hybridisation of the orbitals. It should be compared to the previous table, especially the column labelled "orbital shape". As can be seen from the table the total number of electron pairs (regardless of type) around the central atom determines the hybrid orbitals. The actual shape is determined by how many of these hybrid orbitals are used in bonding to groups.

<u>Finally a few points to remember:</u>

(1) Only sigma bonds are considered in a prediction of geometry.

(2) Sigma bonds result from "head-on" overlap of orbitals.

   eg. s + s

   s + p

(3) Pi bonds result from "sideways" overlap of orbitals, eg.

$p_x + p_x$ or $p_y + p_y$ or $p_z + p_z$

(4) When the bonds in (2) or (3) are normal covalent bonds they involve 2 electrons.

Total Number of Electron Pairs Around Central Atom	Orbital Shape	Hybridisation (bond angle)	Types of Electron Pairs Around Central Atom	Actual Shape
2	linear	sp (180°)	2BP	linear
3	trigonal planar	sp^2 (120°)	3BP	trigonal planar
3	trigonal planar	sp^2	2BP,1LP	angular
4	tetrahedral	sp^3 (109°)	4BP	tetrahedral
4	tetrahedral	sp^3	3BP,1LP	trigonal pyramidal
4	tetrahedral	sp^3	2BP,2LP	angular
5	trigonal bipyramidal	sp^3d (120°, 90°)	5BP	trigonal bipyramidal
5	trigonal bipyramidal	sp^3d	4BP,1LP	sawhorse
5	trigonal bipyramidal	sp^3d	3BP,2LP	bent T
5	trigonal bipyramidal	sp^3d	2BP,3LP	linear
6	octahedral	d^2sp^3 (90°)	6BP	octahedral
6	octahedral	d^2sp^3	5BP,1LP	square pyramidal
6	octahedral	d^2sp^3	4BP,2LP	square planar

3. Molecular Orbitals and Covalent Bonding

The general molecular orbital schemes for homonuclear diatomic molecules are given below and should be memorized.

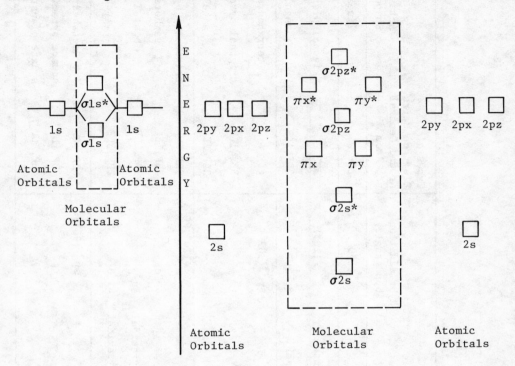

Remember, the molecular orbitals are filled with all the available electrons following the Aufbau principle. The following page contains a number of blank molecular orbital diagrams for your use. Fill these in for the diatomics from Li to Ne. For these diatomics calculate the bond order (B.O.):

$$B.O. = \frac{1}{2} \left[\begin{array}{l} \text{no. electrons in bonding M.O.'s} - \text{no.} \\ \text{of electrons in antibonding M.O.'s} \end{array} \right]$$

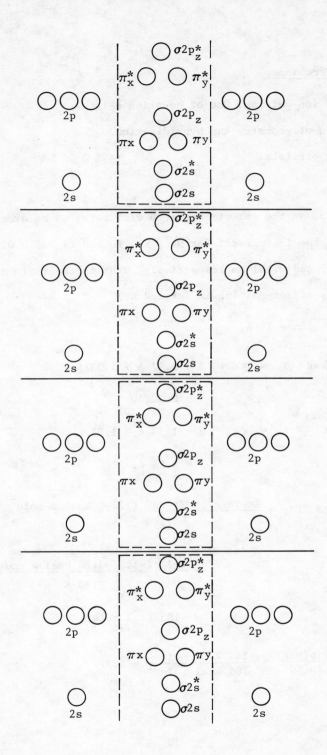

B. Underline{Types of Problems}

1. Energy of ion-pairing. Use of equation 6-1.

2. Prediction of geometry and hybridisation.

3. Molecular orbitals.

Underline{Examples}

1. Question 3. at the end of chapter 6 will serve as an example here. Given the ionic radii for Na^+, K^+, F^- and Cl^- of 0.097, 0.133, 0.133, and 0.181 nm respectively. What is the ratio of the energy of ion-pairing in NaF and KCl?

Note: $1 nm = 10 \, \overset{\circ}{A}$.

$$\text{For NaF} \quad r = r_c + r_a = 0.97 \, \overset{\circ}{A} + 1.33 \, \overset{\circ}{A}$$

$$= 2.30 \, \overset{\circ}{A}.$$

$$\text{For KCl} \quad r = r_c \quad r_a = 1.33 \, \overset{\circ}{A} + 1.81 \, \overset{\circ}{A}$$

$$= 3.14 \, \overset{\circ}{A}.$$

$$\text{For NaF: } E_{ip} = \frac{Z_c \, Z_a \, e^2}{r} = \frac{Z_c \, Z_a}{r \, (\overset{\circ}{A})} \quad (331.0 \text{ kcal } \overset{\circ}{A} \text{ mole}^{-1})$$

$$= \frac{(+1) \, (-1) \, (331.0 \text{ kcal } \overset{\circ}{A} \text{ mole}^{-1})}{2.30 \, \overset{\circ}{A}}$$

$$= -105 \text{ kcal mole}^{-1}$$

$$\therefore \quad \frac{E_{ip}(NaF)}{E_{ip}(KCl)} = \frac{-144 \text{ kcal mole}^{-1}}{-105 \text{ kcal mole}^{-1}} = 1.37.$$

2. For each of the following species

    (a) write Lewis structures including resonance structures

    (b) predict the geometry giving all bond angles

    (c) indicate hybrid orbitals for the central atom
         (underlined)

    (d) indicate the total number of sigma and pi bonds.

    (e) predict whether the species has a dipole moment.

(i) $\underline{B}F_4^-$

    (a) $V = 3 + 7 + 7 + 7 + 7 + 1 = 32$ electrons

$$
\begin{array}{c}
: \overset{\cdot\cdot}{F} : \\
| \\
: \overset{\cdot\cdot}{F} - \overset{\cdot\cdot}{B} - \overset{\cdot\cdot}{F} : \\
| \\
: \overset{\cdot\cdot}{F} : \\
\end{array}
\qquad (32 - 8) = 24 \text{ left} \rightarrow \text{O.K.}
$$

    (b) The total number of electron pairs around boron is 4 and they
         are all bonding electron pairs. Therefore the geometry is
         tetrahedral and the bond angles are 109°. (Use table p.83,84).

    (c) The hybrid orbitals are sp^3 (Use table p.87).

    (d) There are four sigma bonds to boron and no pi bonds.

    (e) The molecule is completely symmetrical with all B-F dipole
         moments exactly cancelling and BF_4^- therefore has no dipole
         moment.

(ii) $\underline{N}O_2^-$

    (a) V = 5 + 6 + 6 + 1 = 18 electrons

$$[:\ddot{O} - \ddot{N} - \ddot{O}:]^-$$

                                                (18 – 4) = 14 left → NO.

$$[:\ddot{O} - \ddot{N} = \overset{..}{\underset{..}{O}}:]^-$$

                                                (18 – 6) = 12 left → YES.

Resonance structures are possible here:

$$[:\ddot{O} - \ddot{N} = \overset{..}{\underset{..}{O}}:]^- \qquad \text{and} \qquad [\overset{..}{\underset{..}{O}} = \ddot{N} - \ddot{O}:]^-$$

    (b) Considering only sigma bonds, the total number of electron pairs around N is <u>3</u>.

                    Total electron pairs = 3

                    Bonding pairs (BP) = 2

                    Lone pairs   (LP) = 1

From the table we see that the shape is angular and the O – N – O angle is 120°.

    (c) Again from the table, for 3 electron pairs around a central atom we have sp^2 hybridisation.

    (d) There are 2 sigma bonds and 1 pi bond.

    (e) Since the molecule is angular there is a dipole moment.

3. Using molecular orbital theory predict the relative bond lengths and heats of dissociation for the species CN^+, CN, and CN^-. Using the molecular orbital diagrams attached one obtains the following configurations for the valence electrons:

CN^+, 8 electrons: σ_{2s}^2 σ_{2s}^{*2} πxy^4

CN , 9 electrons: σ_{2s}^2 σ_{2s}^{*2} πxy^4 σ_{2pz}^1

CN^-, 10 electrons: σ_{2s}^2 σ_{2s}^{*2} πxy^4 σ_{2pz}^2

$\therefore$ the B.O. for $CN^+ = \frac{1}{2} [6 - 2] = 2$

$\quad\quad\quad\quad$ for $CN = \frac{1}{2} [7 - 2] = 2.5$

$\quad\quad\quad\quad$ for $CN^- = \frac{1}{2} [8 - 2] = 3$

The larger the bond order the shorter the bond length and thus the

stronger the bond.

$\quad\quad\quad\quad \therefore$ $\quad$ <u>decreasing bond length</u> $\longrightarrow$

$\quad\quad\quad\quad\quad\quad\quad$ $CN^+ > CN > CN^-$

$\quad\quad\quad$ B.O. $\quad$ 2 $\quad\quad$ 2.5 $\quad$ 3

$\quad\quad\quad\quad\quad\quad$ increasing heat of dissociation $\longrightarrow$

C.   Test Yourself

1.   Given the ionic radii for Ca^{2+}, Ba^{2+}, O^{2-}, and S^{2-} of 0.99, 1.35, 1.40 and 1.84 Å respectively. What is the ratio of the energy of ion-pairing for CaO and BaS? for CaS and BaO?

2.   For each of the following species

   (a) write Lewis structures including resonance structures.

   (b) predict the geometry giving all bond angles.

   (c) indicate the hybrid orbitals used by the central atom (underlined)

   (d) indicate the total number of sigma and pi bonds.

   (e) predict whether the species has a dipole moment.

      (i) $\underline{C}O_2$

      (ii) $\underline{S}O_3$

      (iii) $\underline{C}lO_4^-$

      (iv) $H\underline{C}O_2^-$

      (v) $\underline{S}OCl_2$

      (vi) ClF_5

3. (i)   Using molecular orbital theory, predict whether or not each of the following should be stable to dissociation:

$$O_2^+, \quad Li_2^+, \quad Li_2^{2+}, \quad Ne_2, \quad B_2, \quad Be_2^+, \quad NO^-.$$

   (ii) Predict the relative bond lengths and dissociation energies for the following species using molecular orbital theory.

$$NO, \quad NO^-, \quad NO^+.$$

94

7

The Gaseous, Liquid, and Solid States

A. Points of Importance

1. Ideal Gases.

The ideal gas equation is

$$PV = nRT$$

where   P = pressure

V = volume

n = number of moles

R = gas law constant

The gas law constant can be obtained from standard temperature and pressure conditions. These conditions are commonly referred to as STP and are temperature, 0°C or 273°K and pressure, 1 atm or 760 mmHg. Under these conditions it has been found that <u>one mole</u> of any gas occupies a volume of 22.4 ℓ. This is sometimes referred to as Avogadro's Law and can be stated in another way: at constant temperature and pressure equal volumes of gases contain equal numbers of particles. We can now use the ideal gas equation to calculate a value for R:

$$R = \frac{PV}{nT} = \frac{1 \text{ atm} \times 22.4 \ \ell}{1 \text{ mole} \times 273°K} = 0.0821 \ \frac{\ell \text{ atm}}{\text{mole} \ °K}$$

If a value of $R = 0.0821 \ \dfrac{\ell \text{ atm}}{\text{mole} \cdot °K}$ is used to

solve for any of the remaining variables P,V,n, or T the units corresponding to those of R must be used. It is easiest to use the above R and convert the variables to the correct units rather than try to remember constants with different units.

The ideal gas law equation can be changed so as to calculate density and molecular weight of a gas:

$$PV = nRT$$

But $n = \dfrac{g}{GMW}$ where g = number of grams

GMW = gram molecular weight

$$PV = \frac{g}{GMW} \ RT$$

or $P(GMW) = \dfrac{g}{V} \ RT$

since $\dfrac{g}{V} = \text{density} = d$

$\therefore \ P(GMW) = dRT$

or $d = \dfrac{P(GMW)}{RT}$

and $(GMW) = d \ \dfrac{RT}{P}$

Do not memorize any of these equations. It only takes a few seconds to derive them starting with PV = nRT. Try it.

The equation PV = nRT and any of its derivatives can be used for any set of conditions. For the particular case of STP, Avogadro's Law may also be used. Avogadro's Law (1 mole = 22.4 ℓ) gives two ratios for our use:

$\dfrac{1 \text{ mole}}{22.4 \ \ell}$ and $\dfrac{22.4 \ \ell}{1 \text{ mole}}$ (See Test 7-5).

Another equation that is useful when comparing a gas under different sets of conditions is derived in the text:

$$\frac{P_1 V_1}{n_1 T_1} = \frac{P_2 V_2}{n_2 T_2} \ .$$

In general if we are considering the same gas the number of moles do not change ($n_1 = n_2$) and we have

$$\frac{P_1 V_1}{T_1} = \frac{P_2 V_2}{T_2} \ .$$

When this equation is used with Avogadro's Law we can deal with any problem that can be solved using the ideal gas equation. This will be illustrated in section B.

Chemical reactions involving gases are solved using the same principles developed previously along with PV = nRT. This is clearly illustrated by the sample problem at the end of section 7-3.

Finally when dealing with more than one gas it can be shown that

$P_{total} = P_1 + P_2 + P_3 \ldots P_n$. (n = 1 to ∞). This is called Dalton's Law of Partial Pressures.

97

2. Liquids

Some molecules at the surface of a liquid attain enough energy to overcome the intermolecular forces which hold the liquid together and become gaseous molecules. Some of these gaseous molecules collide with the surface of the liquid and in transferring some of their energy, by collision, to other molecules in the liquid are "re-captured" and become liquid again. These two processes are dynamic and continually occur. In a closed container, figure 7-5, when the rate of molecules escaping from the liquid equals the rate of molecules returning to the liquid the pressure due to the vapor is constant, and the liquid and vapor are said to be in equilibrium. The pressure of the vapor is called the vapor pressure of the liquid. Each individual liquid has its own unique vapor pressure since different liquids have different inter-molecular forces holding them together. The boiling point is that temperature at which a liquid has a vapor pressure equal to that of the atmosphere over it. A normal boiling point is defined at one atmosphere of pressure.

In order to break the forces of attraction in a liquid to convert it to a gas we require energy. This energy is usually expressed as the molar enthalpy of vaporization - the amount of heat that must be added to one mole of a substance to convert it from the liquid to the gaseous state without a temperature change. This quantity, ΔH_{vap} , can be obtained for a given liquid from the slope of the straight line obtained when log (vapor pressure) is plotted against $\frac{1}{T}$ (T in °K).

Such a plot is shown below

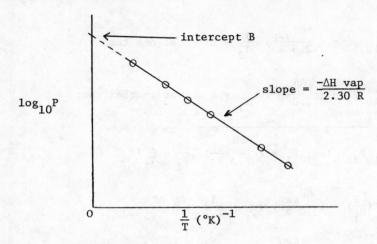

The equation describing the straight line is

$$y = mx + b$$

$$\log_{10} P = (\frac{-\Delta H \text{ vap}}{2.30 \text{ R}}) \frac{1}{T} + B$$

where P is the vapor pressure of the liquid

T is the temperature in °K

R is the gas law constant = 1.99 cal mole^{-1}°K^{-1}

B is a constant characteristic of the liquid.

Thus to calculate ΔH_{vap} we simply need the slope of the line
and use

$$\text{slope} = \frac{-\Delta H_{\text{vap}}.}{2.30 \text{ R}}$$

We can also manipulate the equation for the straight line so that it
can be used to predict the vapor pressure of a liquid at any temperature
provided we know both the vapor pressure at some other temperature and

99

the ΔH_{vap} of the liquid. We proceed in the following manner:

$$\log_{10} P_2 = \left(\frac{-\Delta H_{vap}}{2.3\ R}\right)\ \frac{1}{T_2} + B \qquad \text{at one temp.}$$

$$\log_{10} P_1 = \left(\frac{\Delta H_{vap}}{2.3\ R}\right)\ \frac{1}{T_1} + B \qquad \text{at another temp.}$$

Subtract: $\log_{10} P_2 - \log_{10} P_1 = \left(\frac{-\Delta H_{vap}}{2.3\ R}\right) \left[\frac{1}{T_2} - \frac{1}{T_1}\right]$

$$\log\left(\frac{P_2}{P_1}\right) = \left(\frac{\Delta H_{vap}}{2.3\ R}\right) \left[\frac{1}{T_1} - \frac{1}{T_2}\right]$$

Note: (1) One of the properties of logs is that

$$\log X - \log Y = \log\left(\frac{X}{Y}\right).$$

(2) We took the minus sign from in front of

$\frac{\Delta H_{vap}}{2.3\ R}$ into the brackets giving $\frac{1}{T_1} - \frac{1}{T_2}$.

$\therefore$ we have $\log\left(\frac{P_2}{P_1}\right) = \left(\frac{\Delta H_{vap}}{2.3\ R}\right) \left[\frac{T_2 - T_1}{T_1\ T_2}\right].$

This equation is known as the Clausius-Clapeyron equation.

3. Solids

Solids can be classified according to the units that comprise them, and the type of forces resulting from the particular unit gives the solid its properties. This is summarized in the table on the next page.

100

Classification	Unit	Types of forces
ionic	positive and negative ions in the crystal lattice	large, long-range electrostatic forces over the whole crystal-very strong
metallic	positive ions in a "sea of electrons"	strong metallic bonds
covalent (network)	atoms which share electrons with their neighbors	strong covalent bonds
molecular	molecules interacting with other molecules	(a) polar molecules-weak dipole-dipole forces plus London dispersion forces. (b) nonpolar molecules-weak London dispersion forces. (strongest for large molecules).

The dipole-dipole and London dispersion forces are fully discussed at the beginning of chapter 6. Generally the order of forces in the various compounds is as shown:

covalent bond > metallic > ionic > dipole-dipole > London dispersion

very strong   strong   weak

B.   Types of Problems

1.   Gases

(i)   Find P, V, or T but not given n or g. Use $\dfrac{P_1 V_1}{T_1} = \dfrac{P_2 V_2}{T_2}$

(ii)   Find P, V, or T given g or n.

(iii) Find MW given g, V, T, P.

(iv)   Find density at STP or other T, P.

(v)   Stoichiometry involving gases.

2.   Liquids

(i)   Find heat of vaporization.

(ii)   Find vapor pressure at T_2 given vapor pressure at T_1 and

       given ΔH_{vap}.

3.   Solids

(i)   Predict properties of solids with a knowledge of the types of

       forces involved in the solid.

(ii)   Predict properties of solids, liquids, and gases.

       Examples

1. (i) A 205 ml sample of a gas at 2.00 atm pressure and 25°C is heated to

        125°C without changing the volume. What is the final pressure?

        Since we are not given n or g, use

$$\frac{P_1 V_1}{T_1} = \frac{P_2 V_2}{T_2}$$

        P_1 = 2.00 atm

        P_2 = ?

        In this case $V_1 = V_2$ and the

        equation reduces to $\dfrac{P_1}{T_1} = \dfrac{P_2}{T_2}$

        T_1 = 273 + 25 = 298°K

        T_2 = 273 + 125 = 398°K

        V_1 = 205 ml, V_2 = 205 ml

$$\therefore \quad P_2 = \frac{T_2 P_1}{T_1}$$

$$= \frac{398°K \times 2.00 \text{ atm}}{298°K}$$

$$= 2.67 \text{ atm.}$$

Does this answer make "commonsense"? If you heat a gas in a closed container, the molecules move faster and more strike the walls of the container causing an increase in pressure.

(ii) What volume will be occupied by 9.00 g of CO_2 at 715 mmHg and 25°C?
The <u>easiest way</u> to solve this problem is to use PV = nRT with the appropriate units.

$$P = 715 \text{ mmHg} \times \frac{1 \text{ atm}}{760 \text{ mmHg}} = 0.941 \text{ atm}$$

$$T = 25°C = 273 + 25 = 298°K.$$

$$n_{CO_2} = \frac{9.00 \text{ g}}{44.0 \frac{g}{mole}} = 0.205 \text{ moles.}$$

$$R = 0.0821 \frac{\ell.atm}{mole.°K}$$

$$PV = nRT$$

$$V = \frac{nRT}{P} = \frac{0.205 \text{ mole} \times 0.0821 \text{ } \ell.atm \times 298°K}{0.941 \text{ atm} \qquad mole.°K}$$

$$V = 5.33 \text{ } \ell.$$

There is another method of solving this problem using Avogadro's Law and $\frac{P_1 V_1}{T_1} = \frac{P_2 V_2}{T_2}$.

Let STP be state "1" and the given conditions be state "2". Therefore

103

$$P_1 = 1 \text{ atm} \quad \text{and} \quad P_2 = 0.941 \text{ atm}$$

$$V_1 = ? \qquad \text{and} \quad V_2 = ? \text{ (required)}$$

$$T_1 = 273°K \quad \text{and} \quad T_2 = 298°K.$$

We can calculate V_1 by using Avogadro's Law:

$$V_1 = (\frac{22.4 \text{ } \ell}{\text{mole}} \times 0.205 \text{ mole} = 4.59 \text{ } \ell.$$

$\uparrow$

Avogadro's Law

Now $\quad \dfrac{P_1 V_1}{T_1} = \dfrac{P_2 V_2}{T_2}$

$$V_2 = \frac{P_1 V_1 T_2}{P_2 T_1} = \frac{1 \text{ atm} \times 4.59 \text{ } \ell \times 298°K}{0.941 \text{ atm} \times 273°K}$$

$$V_2 = 5.32 \text{ } \ell.$$

The slightly different answers from the two methods are due to rounding off the volume to 3 significant figures.

(iii) Nicotine has a vapour density of 3.66 g ℓ^{-1} at 1.00 atm and 267°C. Calculate the gram molecular weight of nicotine. <u>Derive</u> the appropriate equation.

$$(GMW) = d \frac{RT}{P}$$

$$= 3.66 \frac{g}{\ell} \times \frac{0.0821\ell \text{ atm} \times 540°K}{\text{mole °K} \times 1 \text{ atm}}$$

$$= 162 \frac{g}{\text{mole}}$$

(iv) Calculate the density of N_2 gas at STP. Since the conditions

are $\underline{STP}$ we will make use of the relation $\underline{22.4\ \ell = 1\ mole}$.

Density requires units of $g\ \ell^{-1}$ ∴ we want the gram molecular weight

of $N_2 = 28.0\ \dfrac{g}{mole}$ ∴ we have $\qquad d = \dfrac{g}{\ell} = \dfrac{28.0\ g\ mole^{-1}}{22.4\ \ell\ mole^{-1}} = 1.25\ g\ \ell^{-1}$.

To calculate the density at any other T and P use the equation

for d derived from PV = nRT.

(v) What volume of $CO_2(g)$ measured at 25°C and 1 atm pressure would be

produced by the combustion of 125 g glucose, $C_6H_{12}O_6(s)$?

First we need a balanced chemical equation:

$$C_6H_{12}O_6(s) + 6O_2(g) \rightarrow 6CO_2(g) + 6H_2O(\ell).$$

Note this is the reverse of photosynthesis.

$\underline{First}$ calculate the number of moles of glucose:

$$n_{glu} = \frac{125\ g}{180\ \dfrac{g}{mole}} = 0.694\ mole.$$

$\underline{From\ the\ equation}$ obtain the number of moles of $CO_2(g)$:

$$0.694\ mole\ glu \times \frac{6\ mole\ CO_2(g)}{1\ mole\ glu} = 4.16\ mole\ CO_2(g).$$

$\underline{Convert}$ 4.16 mole CO_2 to a volume under the specified conditions:

$$PV = nRT$$

$$V = \frac{nRT}{P} = \frac{4.16\ mole \times 0.0821\ \dfrac{\ell\ atm}{mole\ °K} \times 298°K}{1\ atm}$$

$$V = 102\ \ell.$$

2.(i) A plot of log (vapor pressure) versus T^{-1} for bromine gives a straight

line of slope $-1.56 \times 10^3\ °K$. Calculate the heat of vaporization of bromine. We know that slope $= - \dfrac{\Delta H_{vap}}{2.30R}$

$$\therefore \ \Delta H_{vap} = -2.30R \ (\text{slope})$$

$$= -2.30 \times 1.99 \ \text{cal mole}^{-1} \cdot {}^{\circ}K^{-1} \times (-1.56 \times 10^{3} \ {}^{\circ}K)$$

$$= 7.14 \times 10^{3} \ \text{cal mole}^{-1}.$$

$$= 7.14 \ \text{kcal mole}^{-1}.$$

(ii) The normal boiling point of benzaldehyde is 179°C and the ΔH_{vap} = 9.49
kcal mole^{-1}. Estimate the vapor pressure of benzaldehyde at 25°C. This
problem represents a straightforward application of the Clausius –
Clapeyron equation except that you need to know the exact definition
of the normal boiling point: the temperature at which benzaldehyde
has a vapor pressure equal to 760 mmHg or 1 atm.

Tabulate the data:

$$P_1 = ? \qquad\qquad\qquad \text{and } P_2 = 760 \text{ mmHg}$$

$$T_1 = 25° + 273° = 298°K \text{ and } T_2 = 179° + 273° = 452°K$$

The selection of the state "1" or "2" corresponding to 25° or 179°C is
purely arbitrary. The only important point to remember is to make sure
P_1 goes with T_1 and P_2 goes with T_2.

$$\log \left(\frac{P_2}{P_1} \right) = \frac{\Delta H_{vap}}{2.30 \ R} \left[\frac{T_2 - T_1}{T_1 \times T_2} \right]$$

$$\log \left(\frac{760}{P_1} \right) = \frac{9.49 \times 10^{3} \ \text{cal mole}^{-1}}{2.30 \times 1.99 \ \text{cal mole}^{-1} {}^{\circ}K^{-1}} \left[\frac{452-298 \ {}^{\circ}K}{298 \times 452 \ {}^{\circ}K} \right]$$

$$\log \left(\frac{760}{P_1} \right) = 2.37$$

$$\left(\frac{760}{P_1}\right) = \text{antilog } 2.37$$

$$\left(\frac{760}{P_1}\right) = 234$$

$$P_1 = \frac{760 \text{ mmHg}}{234}$$

$$P_1 = 3.25 \text{ mmHg.}$$

3.(i) (a) Predict the relative melting points for the series F_2, Cl_2 Br_2, and I_2. Explain.

The above molecules are classed as molecular and they are all nonpolar. Thus the important forces are London dispersion. These forces increase with the size of the electron clouds and are therefore greatest for I_2 and least for F_2.

$$\therefore \quad \frac{F_2 \ < \ Cl_2 \ < \ Br_2 \ < \ I_2}{\text{melting points increase}} \longrightarrow$$

(b) Predict which species of each of the following pairs has the lower melting point and briefly explain what type of forces must be overcome.

CaO and O_2;

BF_3 and PF_3;

RbCl and BrCl;

SiO_2 and H_2O;

K and H_2.

CaO is ionic and the electrostatic forces are strong. O_2 is nonpolar and only weak London dispersion forces must be overcome to cause melting. Therefore O_2 has the lower melting point.

BF_3 is nonpolar and only London dispersion forces must be overcome. PF_3 is polar and not only has London dispersion forces but also dipole-dipole forces. BF_3 is lower melting.

RbCl is ionic. BrCl has dipole-dipole and London dispersion forces. BrCl is lower melting.

BiO_2 is a covalent species with extremely strong covalent bonds. H_2O is polar and only has relatively weak dipole-dipole and London dispersion forces. H_2O is lower melting.

K has relatively strong metallic bonds whereas H_2 has only weak London dispersion forces to overcome and H_2 has a lower melting point.

(ii) Given the following data for Br_2.

$$\Delta H_{fusion} = 2.6 \text{ kcal mole}^{-1}$$

$$\Delta H_{vap} = 7.3 \text{ kcal mole}^{-1}$$

$$\Delta H_{sublimation} = 9.9 \text{ kcal mole}^{-1}$$

Are the data consistent with your knowledge of the three processes fusion, vaporization, and sublimation? Explain.

The three processes are shown with equations:

fusion: $Br_2(s) \rightarrow Br_2(\ell)$

vaporization: $Br_2(\ell) \rightarrow Br_2(g)$

sublimation: $Br_2(s) \rightarrow Br_2(g)$

108

Sublimation comprises both fusion and vaporization and we can say for a fixed temperature

$$\Delta H_{subl} = \Delta H_{fus} + \Delta H_{vap}$$
$$= 2.6 + 7.3$$
$$= 9.9 \text{ kcal mole}^{-1}.$$

The data are consistent with our knowledge of the three processes.

C. Test Yourself

1. A sample of gas occupies 1.85ℓ at STP. What volume will it occupy at
   $-125°C$ and 0.0100 mmHg pressure?

2. A sample of gas has a volume of 0.500ℓ at 25°C and 50.0 atm pressure.
   What is the pressure of the sample if the volume is changed to 10.0ℓ
   and the temperature is reduced to $-25°C$.

3. What volume will be occupied by 454 g $CH_4(g)$ at 25°C and 1 atm pressure?

4. A 4.00 g sample of gas occupies a volume of 2.12ℓ measured at 125°C
   and 740 mmHg. Calculate the molecular weight of the gas.

5. Calculate the gas law constant in SI units.
$$1\ atm\ =\ 1.013 \times 10^5\ Newton\ meter^{-2}$$
$$1\ liter = 10^{-3}\ meter^3 = 1\ decimeter^3.$$

6. One of the many reactions that occurs due to incomplete combustion of
   gasoline in the automobile is
$$C_8H_{18}(\ell)\ +\ 11O_2(g) \rightarrow 5CO_2(g) + 3CO(g) + 9H_2O(\ell)$$
   Calculate the volume of $CO_2(g)$ and $CO(g)$ formed if measured
   at 25°C and 1 atm, for the combustion of 1 mole $C_8H_{18}(\ell)$.

7. Ozone, O_3, reacts very readily with Ag(s) according to the equation
$$6Ag(s)\ +\ O_3(g)\ \rightarrow\ 3Ag_2O(s)$$
   What volume of air containing 1.0 mg ℓ^{-1} ozone measured at 1 atm
   and 25°C is necessary to react with 1.0 g of Ag(s)?

8. Predict which species of each of the following pairs has the highest
   melting point and state your reasons.

(i) $TiCl_4$ and $TiBr_4$.

(ii)   Ba and F_2.

(iii) Br_2O and Br_2.

(iv)   CaO and CO.

(v)   CO and HI.

(vi) NH_3 and H_2O.

(vii)   Ar and HCl.

9. Predict the relative order of boiling points for the following series:

(i) CH_3Cl, CH_2Cl_2, $CHCl_3$.

(ii)   Ne, Ar, Kr.

10. What is the heat of sublimation, ΔH_{sub}, for acetic acid given $\Delta H_{fusion} =$ 2.8 kcal mole^{-1} and ΔH vap = 7.2 kcal mole^{-1}?

8

An Introduction to Acids and Bases, Chemical Equilibrium, and Chemical Kinetics

A. <u>Points of Importance</u>

1. <u>Lewis Acids and Bases</u>

A <u>Lewis acid</u> is a species that can <u>accept</u> electron density whereas a <u>Lewis base</u> is a species that can <u>donate</u> electron density. An example of a Lewis acid-base reaction is

$$BF_3 + :\overset{..}{\underset{..}{F}}:^- \rightarrow BF_4^-$$

L-acid   L-base   Adduct

In general, Lewis acid-base reactions can be represented by

$$A + :B \rightarrow A:B$$

L-acid   L-base   Adduct

The molecular properties which enhance Lewis acidity and basicity are summarized on the next page. These should be considered to be only guidelines to the categorization of Lewis acids and Lewis bases.

112

Lewis Acid	Lewis Base
High nuclear charge (positive)	low nuclear charge (negative)
large electron affinity	low ionization potential
small size	large size.
(eg.) H^+, Al^{3+}, Si^{4+}	(eg.) OH_2, Br^-, NO_3^-

The special case of a Lewis acid accepting electron density into a vacant antibonding orbital provides the basis of <u>Hydrogen-bonding</u>.

$$\underline{i.e.} \qquad HF: \ + \ HF: \ \rightarrow \ HF \text{------} HF:$$

Since fluorine is highly electronegative, the H bonded to F in HF has a large partial positive charge and thus acts as a reasonably strong Lewis acid. (i.e. the $H^{\delta+}$ models H^+ which is a very good Lewis acid).

2. <u>Brönsted Acids and Bases</u>

A Brönsted acid is a species that <u>donates</u> a proton whereas a <u>Brönsted base</u> is a species that <u>accepts</u> a proton. Brönsted acids are simply protonic Lewis acids (See Eq. (8-12) in the text). An acid-base reaction, then, is one in which proton transfer takes place from one species to another. For example, in the reaction

$$HCl \ + \ H_2O \ \longrightarrow \ \rightleftharpoons \ H_3O^+ \ + \ Cl^-$$

the HCl is acting as a proton donor (∴ is an acid) whereas H_2O is acting as a proton acceptor (∴ a base). The H_3O^+ produced is referred to as the <u>conjugate acid</u> (formed from the base, H_2O) and Cl^- is called the <u>conjugate base</u> (formed from the acid, HCl). The very

113

small arrow to the left indicates that almost all of the HCl has ionized to form H_3O^+ and Cl^- and that there is very little HCl in the solution. This means that HCl is a strong acid. Another example is for the weak acid, acetic acid, CH_3CO_2H, represented simply by HA

$$HA \; + \; H_2O \; \xrightleftharpoons \; H_3O^+ \; + \; A^-$$

$$\text{acid} \qquad \text{base} \qquad \quad \begin{array}{c}\text{conjugate}\\\text{acid}\end{array} \quad \begin{array}{c}\text{conjugate}\\\text{base}\end{array}$$

In this case the small arrow tells us that most of the acetic acid molecules are not dissociated but remain in solution as HA. The **two** arrows, pointing in opposite directions, in both examples mean that these processes are _equilibrium_ processes _i.e._ two forms of the species exists. For HA, in solution we have both HA and A^-. The relative amounts of HA and A^- are given by the value of the equilibrium constant for this reaction.

$$K_a = \frac{[H_3O^+] \; [A^-]}{[HA]} = 1.8 \times 10^{-5}.$$

In the above the subscript "a" means that we are referring to an "acid". The $[H_2O]$ is large and essentially constant and has been incorporated into K_a. One can readily appreciate, from the size of Ka (1.8×10^{-5}), that there are few H_3O^+ and A^- ions in solution and many HA molecules. For the first example, HCl, the K_a value is very large ($\gg 1$) indicating that the numerator of K_a is much larger than the denominator and that the reaction has proceeded far to the right. Thus for a series of acids HA, the size of K_a indicates the strength of the acid in question.

These points can be dramatically demonstrated using the very simple conductivity apparatus shown in figure 3-2 of the text and reproduced below:

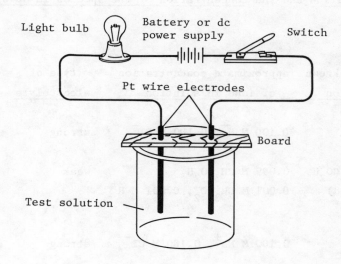

If there are a large number of ions in the test solution the light bulb will glow brightly when the switch is closed. On the other hand if there are relatively few ions in solution electricity will be poorly conducted and the light bulb will only glow faintly.

The table on p. 116 contains the results of experiments performed using this apparatus. Notice the difference between the 0.100 M solutions of HCl, CH_3CO_2H, and sugar. This experiment differentiates between <u>strong</u>, <u>weak</u>, and <u>non electrolytes</u>. Even though <u>in solution</u> the concentration of ions for 0.100 M HCl and 0.100 M CH_3CO_2H is quite different the following example illustrates that the titratable

115

Concentrations of Species in Solution versus Starting Materials

* There is a difference between making up a solution of <u>known</u> <u>concentration</u> and the <u>concentration of the species in solution</u>.

solution of known concentration	approximate concentration of ions in solution	type of electrolyte	"light bulb"
0.100 M HCl	0.100 M H^+, 0.100 M Cl^-	strong	"bright"
0.100 M CH_3CO_2H (acetic acid)	0.099 M CH_3CO_2H, 0.001 M $CH_3CO_2^-$, 0.001 M H^+	weak	"dim"
0.100 NaCl ("salt")	0.100 M Na^+, 0.100 M Cl^-	strong	"bright"
0.100 M NaOH	0.100 Na^+, 0.100 OH^-	strong	"bright"
0.100 M sugar	0.100 M sugar (no ions)	non-electrolyte	did not glow

acidity is the same:

How many ml of 0.100 M NaOH are required to neturalize (titrate) exactly 25.0 ml of 0.100 M HCl and 0.100 M CH_3CO_2H?

The equations describing these neutralization reactions are

$$HCl + NaOH \longrightarrow H_2O + NaCl$$

$$CH_3CO_2H + NaOH \longrightarrow H_2O + NaCH_3CO_2$$

<u>for HCl</u>: n_{HCl} = 0.0250ℓ x 0.100 $\frac{\text{mole}}{\ell}$ HCl = 0.00250 moles

and from the equation, 0.00250 moles HCl x $\frac{1 \text{ mole NaOH}}{1 \text{ mole HCl}}$

$$= 0.00250 \text{ moles NaOH}$$

$\therefore$ since volume x concentration = amount

$\therefore$ volume NaOH required $= \dfrac{0.00250 \text{ moles}}{0.100 \frac{\text{mole}}{\ell}}$

$$= 0.0250 \ \ell$$

$$= 25.0 \ \text{m}\ell \text{ NaOH required}$$

<u>for CH_3CO_2H</u>: $n_{CH_3CO_2H}$ = 0.0250 ℓ x 0.100 $\frac{\text{mole}}{\ell}$ = 0.00250 moles.

0.00250 mole CH_3CO_2H x $\frac{1 \text{ mole NaOH}}{1 \text{ mole } CH_3CO_2H}$ = 0.00250 moles NaOH

$\therefore$ volume required $= \dfrac{0.00250 \text{ moles}}{0.100 \frac{\text{mole}}{\ell}} = 0.0250 \ \ell$
$= 25.0 \ \text{m}\ell$

One can see that the <u>titratable acidity</u> is identical. Remember, there is a difference between <u>stoichiometric concentration</u> and the <u>actual concentration</u> of the species in solution.

117

B. Types of Problems

1. Determine Lewis acids and bases and estimate their relative strengths.

2. Label Bronsted acids and bases and the conjugate acids and bases produced in acid-base reactions.

3. Predict the strengths of acids and bases.

4. Understand the term "titratable acidity"

Examples

1(i) Label the Lewis acid and base in the following reactions

(a) $AlCl_3 + CH_3COCl \rightarrow CH_3CO^+ + AlCl_4^-$

(b) $H_3O^+ + CN^- \rightarrow HCN + H_2O$

(c) $Ag^+ + 2$ pyridine $\rightarrow Ag(pyridine)_2^+$

(a) $\underline{AlCl_3}$ is the L-acid since Al has a vacant orbital to accept a pair of electrons to form $AlCl_4^-$. $\underline{CH_3COCl}$ is the L-base since Cl has a pair of electrons to donate to Al. Note that the product CH_3CO^+ is a L-acid and $AlCl_4^-$ is a L-base.

(b) $\underline{H_3O^+}$ is the L-acid.

$\underline{CN^-}$ is the L-base.

The intermediate complex in this reaction can be thought of as

$$[H-\underset{H}{O}-H \text{---} C\ N].$$

(c) Ag^+ has a vacant orbital and acts as a L-acid.

Pyridine has the structure and thus acts as a L-base

since nitrogen has a lone pair of electrons which are donated.

(ii) Indicate the acidity order expected for increasing strength of

hydrogen bonding toward the L-base trimethylamine, $(CH_3)_3N$.

PH_3, H_2S, HCl

The rule required here is: <u>the more electronegative the atom bonded</u>

<u>to the hydrogen the greater the partial positive charge on hydrogen</u>

<u>and the stronger the L-acid in the hydrogen bonding interaction.</u>

In the above series the electronegativity order is Cl > S > P.

Therefore the strongest hydrogen bond is for HCl, H_2S, and

finally PH_3. The interaction should be viewed as

$$[(CH_3)_3N \text{——} H^{\delta+} Cl^{\delta-}] \qquad \text{for example.}$$

2(i) Write the conjugate bases for the following acids.

(a) HF F^-

(b) $Zn(OH_2)_4^{2+}$ $Zn(OH_2)_3(OH)^+$

(c) OH_2 OH^-

(a) The <u>conjugate base</u> is the same compound <u>minus</u> a proton i.e. the

result produced after the acid has donated its proton.

Therefore the conjugate base of HF is F^-.

(c)   The conjugate base of $Zn(OH_2)_4^{2+}$ is

$$Zn(OH_2)_3(OH)^+.$$

The conjugate base of OH_2 is OH^-.

(ii) In the following reactions label the Brönsted acids and bases as well as the conjugate acids and bases.

(a) $CN^- + H_3O^+ \rightleftharpoons HCN + H_2O$

(b) $CN^- + H_2O \rightleftharpoons HCN + OH^-$

(c) $NH_4^+ + H_2O \rightleftharpoons NH_3 + H_3O^+$

(d) $HCO_3^- + OH^- \rightleftharpoons H_2O + CO_3^{2-}$

(e) $HCO_3^- + H_2O \rightleftharpoons H_2CO_3 + OH^-$

(a) $CN^- + H_3O^+ \rightleftharpoons HCN + H_2O$
base   acid   conjugate   conjugate
                    acid   base

The $\underline{CN^-}$ $\underline{accepts}$ a proton from H_3O^+ and becomes HCN. Therefore CN^- is a $\underline{base}$ and produces a $\underline{conjugate\ acid}$. The $\underline{H_3O^+}$ $\underline{donates}$ a proton to CN^- and becomes H_2O. Therefore H_3O^+ is an $\underline{acid}$ and produces a $\underline{conjugate\ base}$.

(b) $CN^- + H_2O \rightleftharpoons HCN + OH^-$
base   acid   conjugate   conjugate
                    acid   base

(c) $NH_4^+ + H_2O \rightleftharpoons NH_3 + H_3O^+$
acid   base   conjugate   conjugate
                    base   acid

120

(d) HCO_3^- + OH^- $\rightleftharpoons$ H_2O + CO_3^{2-}

     acid   base   conjugate   conjugate
                                      acid   base

(e) HCO_3^- + H_2O $\rightleftharpoons$ H_2CO_3 + OH^-

    base   acid   conjugate   conjugate
                                    acid   base

It is clear from (d) and (e) above that HCO_3^- can act both as an acid and a base depending on the reaction. Which way it acts is easily determined by using the Brönsted definitions.

3. In each of the following pairs, choose the species that would be the stronger acid:

(a) HBrO or HClO.

(b) $HClO_2$ or HClO.

(c) H_3PO_4 or $H_2PO_4^-$.

In terms of our general compound <u>Y − O − H</u>, the acid strength increases as Y becomes more electron withdrawing or more electronegative. Attaching electronegative substituents to Y (such as O) makes Y more electronegative.

(a) We can rewrite HBrO and HClO in the form <u>Y−O−H</u>:

<u>Br−O−H   and   Cl−O−H.</u>

Since Cl is more electronegative than Br, Cl produces a larger partial positive charge on the hydrogen and makes it easier to remove when attacked by water. Therefore HClO is a stronger acid than HBrO.

121

(b) Again rewriting as O–Cl–O–H and Cl–O–H we can see that the Cl of O–Cl–O–H is more electronegative than the Cl of Cl–O–H because of the extra attached O. Therefore $HClO_2$ is more acidic than $HClO$.

(c) The two acids are similar

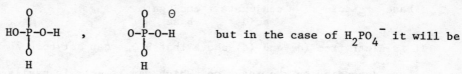

but in the case of $H_2PO_4^-$ it will be more difficult to separate a H^+ ion from a species that is already negatively charged. In H_3PO_4 we are separating a H^+ ion from a neutral species. Therefore H_3PO_4 is more acidic than $H_2PO_4^-$.

4.   The K_a's for some weak acids are given below:

weak acid	K_a	
acetic acid, CH_3CO_2H	1.8×10^{-5}	2
hydrofluoric acid, HF	7.0×10^{-4}	1
hypochlorous acid, HClO	3.2×10^{-8}	3
hydrocyanic acid, HCN	4.0×10^{-10}	4

(a) For 0.100 M solutions of these acids predict the order of H^+ concentrations.

(b) For 25.0 ml of each acid in (a), which would require the most 0.100 M NaOH to neutralize the acids?

(a) The K_a for the general acid HA is

$$K_a = \frac{[H_3O^+][A^-]}{[HA]}$$ which comes from the equation

$$HA + H_2O \rightleftharpoons H_3O^+ + A^-$$

Thus for equal concentration of HA the magnitude of K_a tells us the relative amount of H_3O^+ in solution. i.e. a large value of K_a means a large number of H_3O^+ and A^- ions in the solution. A small number for K_a means very few H_3O^+ and A^- ions in the solution.*

Thus for the given solutions HF will give the most H^+ ions in solution whereas HCN will give the least number of H^+ ions. The relative ordering is

$$HF > CH_3CO_2H > HClO > HCN$$
$$K_a = 7.0 \times 10^{-4} \quad 1.8 \times 10^{-5} \quad 3.2 \times 10^{-8} \quad 4.0 \times 10^{-10}$$

(b) Since all solutions have the same concentration of acid, HA, they all have the same amount of titratable acidity and therefore require the same amount of 0.100 M NaOH for neutralization. Write the equations and do the calculations.

* Generally we will indicate an equilibrium with double arrows of equal length and obtain information as to the extent of the reaction from the magnitude of the equilibrium constant.

C. Test Yourself.

1. Which of the following species can **act** as a Brönsted acid or base, or a Lewis acid or base? Write an equation to justify your answer.

   (i) H_2O.

   (ii) BF_3.

   (iii) $Al(OH_2)_6^{3+}$.

   (iv) pyridine, .

2. Write equations for the following species behaving as Brönsted bases toward water.

   (i) $:N(CH_3)_3$. $+ H_2O \rightarrow HN(CH_3)_3^+ + OH^-$

   (ii) CN^-. $+ H_2O \rightarrow HCN + OH^-$

   (iii) F^-. $+ H_2O \rightarrow HF + OH^-$

3. Write equations to illustrate the reaction of water as a Brönsted acid and a Brönsted base. Satisfy yourself that water is also acting as a Lewis acid and a Lewis base in these reactions.

4. In terms of the size of K_a explain why the conjugate base of a strong acid is weak and why the conjugate base of a very weak acid is reasonably strong. Use equations.

5. Define a Brönsted conjugate base and a Brönsted conjugate acid and from the following list pick out the Brönsted acid – conjugate base and Brönsted base – conjugate acid pairs: H_2O, CO_3^{2-}, OH^-, H_2CO_3, OH_3^+, HCO_3^-, O^{2-}, HCl, Cl^-.

124

6. For the following acids give the conjugate bases and list them in order of decreasing strength: $HClO_4$, CH_3CO_2H, HCN, HCl, HNO_2.

9

Water: Properties and Reactions.
Pollution and Purification.
Colligative Properties

A. Points of Importance

1. The reactivity of water – hydration and hydrolysis.

A question very frequently asked by freshman chemistry students is "How can I tell which acids and bases are strong, which are weak, and which "salts", C^+X^-, give acidic, basic or neutral solutions"? At the beginning this is no easy task so a few tips on "what to" and "what not to" memorize are in order.

(a) Memorize the strong acids and bases commonly encountered in general chemistry – the rest are almost always weak acids and bases.

Strong Acids   Strong Bases

HCl, HBr, HI   NaOH, KOH, LiOH, RbOH, CsOH

$HClO_4$ $Ba(OH)_2$, $Ca(OH)_2$, $Sr(OH)_2$.

HNO_3

H_2SO_4

Remember these "strong" species are characterized by complete ionization

of the H^+ or OH^- in the compound.

(b) Weak acids are characterized by the fact that they do not completely

ionize in solution <u>i.e.</u> they react in solution to produce H_3O^+ to only

a small extent.

Some commonly encountered weak acids are listed below:

<u>Weak Acids</u>

HCN, HF, H_2S, H_2CO_3, HNO_2, CH_3CO_2H (acetic),

$C_2H_5CO_2H$ (propionic), $C_6H_5CO_2H$ (benzoic),

HCO_2H (formic).

(c) For any one of the weak acids in (b), say formic acid, the following

equilibrium can be written:

$$HCO_2H + H_2O \rightleftharpoons H_3O^+ + HCO_2^-$$

or, in general for the weak acid HA,

$$HA + H_2O \rightleftharpoons H_3O^+ + A^-.$$ Remember the symbol

K_a represents the equilibrium constant for this reaction. In the above

equations only a small amount of H_3O^+ is produced and thus the acid is

weak. In the above, A^- is the conjugate base of the weak acid HA and

since the equilibrium lies far to the left (i.e. almost all HA in

solution; very little H_3O^+ and A^-) one can see that A^- wishes to be

coordinated to form HA. Thus if A^- is put into solution via the C^+A^-

compound then A^- reacts with H_2O to form HA as shown:

$$A^- + H_2O \rightleftharpoons HA + OH^-.$$

Since this reaction produces OH^-, A^- is considered to be a weak base

127

just as HA is considered to be a weak acid.

Therefore one can state that the conjugate base, A⁻, of any weak acid, HA, acts as a weak base when obtained in solution from a salt C^+A^-. The action of A⁻ on H_2O is referred to as hydrolysis.

(d) If we are dealing with a strong acid, say HCl, then the equilibrium lie far to the right:

$$HCl \ + \ H_2O \ \rightleftharpoons \ H_3O^+ \ + \ Cl^-$$

Therefore if Cl⁻ is present in solution from the salt C^+Cl^- there is no tendency to react with H_2O as the conjugate base of a weak acid does. This is simply a consequence of the fact that HCl is a strong acid. In other words, any conjugate base of a strong acid does not cause hydrol of water to occur. Memorize the fact that Cl⁻ is neutral in aqueous solution.

(e) There are also a number of compounds which are neutral and behave as weak bases in exactly the same manner as the conjugate bases of weak acids. Some of these are listed below:

neutral weak bases

NH_3

$(CH_3)NH_2$

$(CH_3)_2NH$

$(CH_3)_3N$

C_5H_5N (pyridine)

$C_6H_5NH_2$ (aniline), ⬡—NH_2

These neutral weak bases cause hydrolysis of water as illustrated for

128

methylamine:

$$(CH_3)NH_2 + H_2O \rightleftharpoons (CH_3)NH_3^+ + OH^-.$$

Notice the similarity of the above equation to

$$A^- + H_2O \rightleftharpoons HA + OH^-.$$

Methylamine thus causes hydrolysis of water to form OH^- and $(CH_3)NH_3^+$ (the conjugate _acid_ of the weak base $(CH_3)NH_2$). One can readily appreciate from the previous analysis as well as the last two equations that $(CH_3)NH_3^+$ in solution must behave as an _acid_. Therefore if we put $(CH_3)NH_3^+$ in solution via the salt, say $(CH_3)NH_3^+Cl^-$, we have an acidic solution because of the equation

$$(CH_3)NH_3^+ + H_2O \rightleftharpoons (CH_3)NH_2 + H_3O^+.$$

You can see that this equation is in turn similar to

$$HA + H_2O \rightleftharpoons A^- + H_3O^+.$$

Memorize the fact that NH_4^+ is a weak acid.

(f) In summary, two equations describe the majority of cases you will meet:

$$HA + H_2O \rightleftharpoons H_3O^+ + A^- \longrightarrow K_a$$
$$(\text{or } BH^+ + H_2O \rightleftharpoons H_3O^+ + B \longrightarrow K_h)$$
$$A^- + H_2O \rightleftharpoons HA + OH^- \longrightarrow K_h$$
$$(\text{or } B + H_2O \rightleftharpoons BH^+ + OH^- \longrightarrow K_b)$$

where HA can be for eg. HCN or NH_4^+ and A^- can be for eg. CN^- or NH_3. You can well appreciate that for C^+X^- compounds the properties of both C^+ and X^- must be considered in order to predict whether a solution of C^+X^- will be acidic, basic or neutral.

Now we are in a position to make our own table of neutral, acidic, and basic species.

Neutral	Acidic	Basic
Cl^-, Br^-, I^-	HSO_4^-	CN^-, F^-, $CH_3CO_2^-$
ClO_4^-	$H_2PO_4^-$	HS^-, S^{2-}
NO_3^-	NH_4^+	HCO_3^-, CO_3^{2-}
alkali metal cations, M^+	Al^{3+}, Cr^{3+}	PO_4^{3-}, HPO_4^{2-}
Ba^{2+}, Ca^{2+}, Sr^{2+}	Be^{2+}, Fe^{2+}	SO_4^{2-}

Note: (1) In the neutral column all the anions are the conjugate bases
 of strong acids and all the cations come from strong bases.

(2) In the basic column all the anions are conjugate bases of
 weak acids.

2. Colligative Properties

Colligative properties are those that depend on the number of solute
particles in a solution and not on the type of particle. These propert
are vapor pressure lowering, boiling point elevation, freezing point
depression, and osmotic pressure.

Vapor Pressure Lowering

When a solute is dissolved in a pure solvent the vapor pressure
above the solution is less than for the pure solvent alone. The vapor
pressure lowering, ΔP v.p., is given by

$$\Delta P \text{ v.p.} = X_2 P_1^\circ$$

where X_2 is the mole fraction of the solute in solution, (moles solute
total number moles) and P_1° is the vapor pressure of the pure solvent.

130

Boiling Point Elevation and Freezing Point Depression

When a solute is dissolved in a pure solvent the solution has a higher boiling point than the boiling point of the pure solvent <u>and</u> the solution has a lower freezing point than the pure solvent. This can be expressed by the two equations below:

$$\Delta T_{bp} = k_{bp} \, xm$$

$$\Delta T_{fp} = k_{fp} \, xm$$

where ΔT_{bp} is the boiling point elevation in °C

k_{bp} is the molal boiling point elelvation constant

m is the molality of the solute in the solution and

ΔT_{fp} is the freezing point depression in °C

k_{fp} is the molal freezing point constant

m is the molality of the solute in the solution.

The molality is another way of expressing concentration with the special feature that it is temperature independent. The molality, m, is defined as

$$m = \frac{\text{number of moles of solute}}{1000 \text{ g solvent}}$$

Osmotic Pressure

<u>Osmosis</u> is the process where solvent molecules diffuse through a semipermeable membrane (does not permit passage of solute molecules) to a region of higher solute concentration. The pressure applied to the region of higher solute concentration that just prevents osmosis is

131

called the osmotic pressure. This pressure is equivalent to that exerted

by the height of the column in figure 9-10(a). If more than the osmotic

pressure is applied then solvent molecules are forced from the region of

high solute concentration to the region of low solute concentration. This

is called reverse osmosis. For example if we separated sea water and pure

water by a semipermeable membrane osmosis would occur and pure water

would diffuse into the sea water. If we apply the osmotic pressure

of the sea water to the sea water solution osmosis will be prevented.

Now, if we apply a greater pressure we can force pure water to diffuse

from the sea water through the membrane into the pure water. This

latter process, called reverse osmosis, is one of the means of

obtaining pure water from sea water.

B. Types of Problems

1. Reactions of and in water.

2. Colligative Properties

    (i)   Vapor pressure lowering.

    (ii) Boiling point elevation and freezing point depression.

    (iii) Osmotic pressure.

Examples

1.(i) Illustrate the difference between the terms hydration and

       hydrolysis for the situation of $CrCl_3$ dissolving in water

       using net ionic equations.

  (ii) Predict whether the following 0.1 M solutions will be acidic,

        basic, or neutral and justify with a net ionic equation:

                (a) NaF.

                (b) C_5H_5NHCl (pyridinium chloride)

                (c) NH_4HCO_3

  (iii) Questions 17 and 18 at the end of chapter 9.

        For each of the following indicate whether or not an aqueous solution

        can tolerate large concentrations of the two ions simultaneously and

        if it cannot, write a balanced equation for the reaction that will

        occur. For those systems in which large concentrations of the ions

        can simultaneously exist, which will undergo hydrolysis?

    (a) H^+ and CN^-.

    (b) Na^+ and CN^-.

133

(c) H^+ and NO_3^-.

(d) Al^{3+} and OH^-.

(e) NH_4^+ and Cl^-.

(f) NH_4^+ and OH^-.

(g) Ba^{2+} and SO_4^{2-}.

(h) H^+ and CO_3^{2-}.

(i) $C_5H_5NH^+$ and OH^-.

(j) $C_5H_5NH^+$ and Cl^-.

(iv) Classify the solutes listed below according to the flow chart on p290 of the text: NH_4Cl, HF, $CsOH$, HCl, HNO_3, NH_3, $(CH_3)_3N$, $NaCN$, $Cr(ClO_4)_3$.

1.(i) When $CrCl_3$ dissolves in solution we have Cr^{3+} and $3Cl^-$ ions. For Cr^{3+} we first have <u>hydration</u>:

$$Cr^{3+} + 6H_2O \longrightarrow Cr(OH_2)_6^{3+}$$

<u>i.e.</u> water molecules surround the cation. The hydrated species $Cr(OH_2)_6^{3+}$ can undergo <u>hydrolysis</u>:

$$Cr(OH_2)_6^{3+} + H_2O \Longleftrightarrow Cr(OH_2)_5(OH)^{2+} + H_3O^+$$

<u>i.e.</u> give up a proton to H_2O.

In this case Cl^- is also <u>hydrated</u> (surrounded by water molecules) but the number of water molecules is not specified.

(ii)(a) NaF dissolved in water gives a <u>basic</u> solution. Remember, Na^+ has no tendency to cause hydrolysis of water since it is an alkali metal cation and comes from the strong base NaOH. Thus Na^+ undergoes hydration <u>only</u>. On the other hand, F^- is the conjugate <u>base</u> of the

134

weak acid HF and as such must act as a base in causing hydrolysis of water:

$$F^- + H_2O \rightleftharpoons HF + OH^-.$$

(b) C_5H_5NHCl dissolved in water gives an _acidic_ solution. Again, Cl^- is hydrated but comes from the strong acid HCl and therefore has no tendency to hydrolyze water. $C_5H_5NH^+$ is the conjugate _acid_ of the base pyridine and as such acts as an acid:

$$C_5H_5NH^+ + H_2O \rightleftharpoons C_5H_5N + H_3O^+.$$

(c) For NH_4HCO_3 in solution the two ions we must consider are NH_4^+ and HCO_3^-. You will immediately recognize that NH_4^+ is a weak acid:

$$NH_4^+ + H_2O \rightleftharpoons NH_3 + H_3O^+.$$

However, HCO_3^- is the conjugate base of the weak acid, H_2CO_3, and acts as a base in causing the hydrolysis of water:

$$HCO_3^- + H_2O \rightleftharpoons H_2CO_3 + OH^-.$$

Thus we need to know to what extent these two reactions proceed before we can predict whether the solution is acidic, basic, or neutral. _i.e._ the magnitude of the equilibrium constants K_{h_1} (for NH_4^+) and K_{h_2} (for HCO_3^-) will tell us the relative number of H_3O^+ (from NH_4^+) and OH^- (from HCO_3^-) ions in solution. If $K_{h_1} > K_{h_2}$ then the solution will be acidic. If $K_{h_2} > K_{h_1}$ then the solution will be basic. And if

$K_{h_1} = K_{h_2}$ the solution will be neutral. In this case $K_{h_1} =$ 5.6 x 10^{-10} and $K_{h_2} = 2.4$ x 10^{-8} and therefore the solution is _basic_.

(iii) (a) $\underline{H^+ \text{ and } CN^-}$. Since HCN is a weak acid an aqueous solution cannot tolerate large concentrations of H^+ and CN^- ions. The reaction that occurs is $H^+ + CN^- \rightleftharpoons HCN$.

(b) $\underline{Na^+ \text{ and } CN^-}$. Since NaCN is a salt it is completely ionized in solution and large concentrations of Na^+ and CN^- can be tolerated. In this case Na^+ simply undergoes _hydration_ but CN^- being the conjugate base of the weak acid HCN causes _hydrolysis_ of water:

$$CN^- + H_2O \rightleftharpoons HCN + OH^-.$$ The solution is therefore basic.

(c) $\underline{H^+ \text{ and } NO_3^-}$. Since HNO_3 is a strong acid, an aqueous solution can tolerate large concentrations of H^+ and NO_3^-. The solution is, of course acidic and hydrolysis does not occur.

(d) $\underline{Al^{3+} \text{ and } OH^-}$. In aqueous solution Al^{3+} and OH^- react to first form $Al(OH)_3(s)$. If further OH^- is added the following reaction occurs

$$Al(OH)_3(s) + 3OH^- \rightarrow Al(OH)_6^{3-}.$$

Thus Al^{3+} and OH^- cannot exist simultaneously in aqueous solution.

(e) $\underline{NH_4^+ \text{ and } Cl^-}$. NH_4Cl is a salt (strong electrolyte) and is completely ionized in solution and large amounts of NH_4^+

and Cl^- ions can coexist in solution. Cl^- is a neutral species and NH_4^+ is weak acid:

$$NH_4^+ + H_2O \rightleftharpoons NH_3 + H_3O^+.$$

(f) $\underline{NH_4^+ \text{ and } OH^-}$. Large concentrations of OH^- and NH_4^+ cannot exist in solution because the reaction

1) $NH_4^+ + OH^- \rightleftharpoons NH_3 + H_2O$ has a large equilibrium constant. This can be appreciated by the fact that NH_3 is a $\underline{weak}$ base and the reaction

2) $NH_3 + H_2O \rightleftharpoons NH_4^+ + OH^-$ has a $\underline{small}$ equilibrium constant.

For reaction 2)

$$K_2 = \frac{[NH_4^+]\,[OH^-]}{[NH_3]}.$$

For reaction 1)

$$K_1 = \frac{[NH_3]}{[NH_4^+]\,[OH^-]}. \text{ One can readily see that}$$

$$K_2 = \frac{1}{K_1}. \quad \text{Therefore if } K_2 \text{ is small, then } K_1 \text{ must be large as argued above.}$$

It is always true that the equilibrium constant for a reaction written in one direction, $K+$, is equal to the reciprocal of the equilibrium constant written in the opposite direction, $K-$.

(eg.) $\quad A^- + H_2O \rightleftharpoons HA + OH^- \quad K+$

$\qquad HA + OH^- \rightleftharpoons A^- + H_2O \quad K-$

$$\therefore \quad K+ = \frac{1}{K-}. \quad \text{Verify this!}$$

137

(g) $\underline{Ba^{2+} \text{ and } SO_4^{2-}}$. These ions cannot exist simultaneously in
solution because of the following reaction:
$$Ba^{2+} + SO_4^{2-} \longrightarrow BaSO_4(s).$$

(h) $\underline{H^+ \text{ and } CO_3^{2-}}$. Large concentrations of these ions cannot coexist
in solution since the reaction below occurs:
$$H^+ + CO_3^{2-} \longrightarrow HCO_3^-.$$ This can be understood by noting
that HCO_3^- does not act as an acid but as a base. In addition
CO_3^{2-} is a weak base that causes hydrolysis of water and therefore
reacts with acids readily.

(i) $\underline{C_5H_5NH^+ \text{ and } OH^-}$. Large concentrations of these ions cannot coexist in aqueous solution because the following reaction takes
place $\quad C_5H_5NH^+ + OH^- \rightleftharpoons C_5H_5N + H_2O.$
This can readily be appreciated by realizing that C_5H_5N (pyridine)
is a weak base and as such the reaction below
$$C_5H_5N + H_2O \rightleftharpoons C_5H_5NH^+ + OH^-$$
does not proceed very far to the right.

(j) $\underline{C_5H_5NH^+ + Cl^-}$. This is simply a strong electrolyte (C_5H_5NHCl)
and large concentrations of the ions do exist in solution. Cl^-
is neutral and $C_5H_5NH^+$ is acidic because of the reaction
$$C_5H_5NH^+ + H_2O \rightleftharpoons C_5N_5N + H_3O^+.$$

(iv) NH_4Cl, HF, HCl, $CsOH$, HNO_3, NH_3, $(CH_3)_3N$, $NaCN$, $Cr(ClO_4)_3$.

Does the solute contain ionizable H^+ or OH^-?

Yes ↙   ↘ No

HF, HCl, CsOH, HNO_3, NH_4Cl NH_3, $(CH_3)_3N$, NaCN, $Cr(ClO_4)_3$

Is it an HX, BH^+, or MOH type of compound?

Is it a neutral base or a C^+X^- compound?

1) **HX type:** HF, HCl

$$HF + H_2O \rightleftharpoons H_3O^+ + F^-$$
$$HCl + H_2O \rightleftharpoons H_3O^+ + Cl^-$$

2) **BH^+ type:** $NH_4^+Cl^-$

$$NH_4^+ + H_2O \rightleftharpoons NH_3 + H_3O^+$$

The anion must also be considered:
here Cl^- is the conjugate base of
a strong acid and is thus neutral.
If we were given $NH_4^+F^-$ the situation
would have been more complicated
and the acidity would depend on the
magnitude of $K_h(NH_4^+)$ and $K_h(F^-)$.

3) **MOH type:** CsOH, HNO_3 ($\overset{O-H}{\underset{O\;\;O}{N}}$).

$$CsOH + 6H_2O \rightleftharpoons Cs(OH_2)_6^+ + OH^-$$

$$HNO_3 + H_2O \rightleftharpoons H_3O^+ + NO_3^-$$

1) **Neutral base:** NH_3, $(CH_3)_3N$.

$$NH_3 + H_2O \rightleftharpoons NH_4^+ + OH^-$$
$$(CH_3)_3N + H_2O \rightleftharpoons (CH_3)_3NH^+ + OH^-$$

2) **C^+X^- compound:** NaCN, $Cr(ClO_4)_3$.

Consider C^+ and A^- separately

CATIONS

1) Small and highly charged: Cr^{3+}
$$Cr^{3+} + 6H_2O \rightleftharpoons Cr(OH_2)_6^{3+}$$
$$Cr(OH_2)_6^{3+} + H_2O \rightleftharpoons Cr(OH_2)_5(OH)^{2+} + H_3O^+$$

2) Large and low charge: Na^+
$$Na^+ + 6H_2O \rightleftharpoons Na(H_2O)_6^+$$

ANIONS

1. Conjugate base of a weak acid:
CN^- is basic.
$$CN^- + H_2O \rightleftharpoons HCN + OH^-$$

2. Conjugate base of strong acid:
ClO_4^- is neutral.

2. (i)   Calculate the vapor pressure lowering at 100°C for an aqueous

 solution prepared by dissolving 100 g of ethylene glycol

 $C_2H_4(OH)_2$, in 500 g water. (the vapor pressure of water at 100°C

 is 760 mm Hg).

(ii)   Question 26 at the end of chapter 9.

 A nonelectrolyte weighing 1.125 g was added to 10.0 g of benzene.

 This benzene solution froze at 2.92°C. Calculate the molecular

 weight of the solute.

(i)   The vapor pressure lowering is given by the equation

$$\Delta Pv.p. = X_2 P_1^{\,o}$$

$$X_2 = \frac{\text{number moles ethylene glycol}}{\text{moles ethylene glycol + moles water}}$$

$$= \frac{100 \text{ g} / 62 \text{ g mole}^{-1}}{\dfrac{100 \text{ g}}{62 \text{ g mole}^{-1}} + \dfrac{500 \text{ g}}{18 \text{ g mole}^{-1}}}$$

$$= \frac{1.61}{1.61 + 27.8} = 0.055$$

$$\Delta Pv.p. = 0.055 \times 760$$

$$= 42 \text{ mmHg.}$$

(ii)   Since benzene freezes at 5.48°C, the freezing point lowering is

 5.48 − 2.92 = 2.56°C.

 Since $\Delta T_{f.p.} = k_{fp} \times m$ (From Table 9-3 in text, $k_{fp} = 5.12°Cm^{-1}$)

$$m = \frac{2.56°C}{5.12°Cm^{-1}} = 0.500 \text{ m}$$

Therefore we require 0.500 moles of solute per kg benzene. However we have 1.125 g solute in 10.0 g benzene i.e. $\dfrac{1.125 \text{ g solute}}{10.0 \text{ g benzene}}$.

Therefore for 1000 g benzene we have

$$\dfrac{1.125 \text{ g solute}}{10.0 \text{ g benzene}} \times 1000 \text{ g benzene} = 112.5 \text{ g solute.}$$

This 112.5 g solute must give a 0.500 m solution.

Therefore we have $\dfrac{112.5 \text{ g solute}}{1000 \text{ g benzene}} \times \dfrac{1000 \text{ g benzene}}{0.500 \text{ moles solute}}$

$$= 225 \ \dfrac{g}{\text{mole}} = \text{gram molecular weight}$$

C. Test Yourself

1. Predict whether the following 0.1 M solutions will be acidic, basic, or neutral and justify your answer using a net ionic equation.

    (a) NaH_2PO_4.

    (b) $Fe(ClO_4)_2$.

    (c) $NH_4(CH_3CO_2)$.

    (d) $RbNO_3$.

    (e) NaHS.

2. Why is it misleading to refer to a concentrated solution of aqueous ammonia, NH_3, as ammonium hydroxide solution, NH_4OH?

3. Illustrate all the processes which occur when the following are dissolved in water by writing equations. Also name the processes.

    (a) $Ba(ClO_4)_2$.

    (b) $FeCl_2$.

    (c) $AlCl_3$.

4. For the following reactions show that $K = K_h^{-1}$.

    $$F^- + H_2O \rightleftharpoons HF + OH^- \qquad K_h$$
    $$HF + OH^- \rightleftharpoons F^- + H_2O \qquad K$$

5. Calculate the vapor pressure lowering for a solution prepared by dissolving 10.0 g naphthalene, $C_{10}H_8$, in 100 g of benzene, C_6H_6, at 25°C. The vapor pressure of pure benzene at 25° is 97.0 mmHg.

6. What is the boiling point of the solution prepared by dissolving 10.0 g naphthalene in 100 g benzene?

7. Addition of 1.50 grams of caffeine to 25.0 grams of benzene results in a solution that freezes at 3.89°C. Calculate the molecular weight of caffeine.

8. Explain how you would use osmosis to obtain pure water from sea water.

9. Classify the following solutes according to the flow chart on p 290 of the text: C_5H_5NHCl, HCN, KOH, $HClO_4$, KF, C_5H_5N, $Fe(ClO_4)_2$.

10

Periodic Trends in the Descriptive Chemistry of Elements and Compounds. Oxidation-Reduction

A.   Points of Importance

1.   Periodic Trends.

The basic periodic trends were outlined in chapter 6 and can be extended to trends in melting point and enthalpy of atomization on a limited basis. For the metallic elements on the left hand side of the periodic table the melting points and heats of atomization decrease as one goes down a group. As one moves from left to right across the table the metallic character decreases and the properties of the elements change drastically. Some of these properties are outlined on the **two blank** periodic tables on p. 145. On the third blank periodic table the trends in electronegativity are again repeated. The variations in electronegativity are important in considering trends in the re-activity of hydrogen and hydrides. It is the difference in electro-negativity between the central atom and hydrogen, for example, which

MELTING POINT AND ENTHALPY OF ATOMIZATION

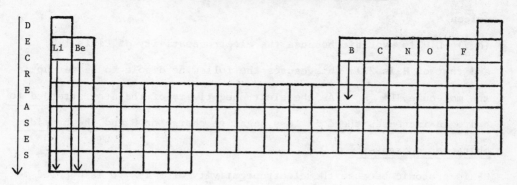

SPECIAL PROPERTIES

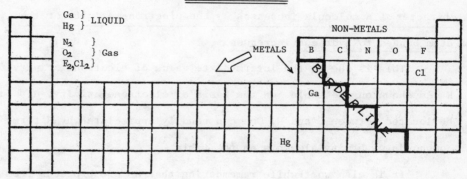

ELECTRONEGATIVITY INCREASES

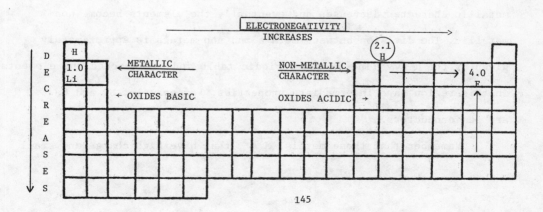

145

determines the hydridic character or the protonic character of a molecule.

(eg.) LiH is hydridic because the electronegativity of Li is 1.0 and that of H is 2.1. This causes the following dipole to be set up in the molecule, $Li^{\delta+} H^{\delta-}$. As the electronegativity of the atom attached to hydrogen increases there is less negative charge on H and the hydridic character decreases.

HF is protonic because the electronegativity of F is 4.0 as opposed to 2.1 for H. The following dipole is set up $H^{\delta+} F^{\delta-}$. The protonic character of a molecule increases as the electronegativity of the atom bonded to hydrogen increases.

This is due to the intermediate value of electronegativity for H. As shown on the table, on the basis of electronegativity, H should be located between C and N. One can readily appreciate why H forms compounds with many elements on the table.

It is also worthwhile remembering that as one proceeds from left to right on the periodic table (increasing electronegativity) the metallic character decreases and eventually the elements become non-metallic. The division between metals and non-metals is approximately given by the heavy line on the periodic table and as expected the elements in this region have intermediate properties. For example, Si and Ge are semiconductors.

Remember that those metals, $M^+ A^-$, that have high charge and small

146

size cause hydrolysis of water whereas those with small charge and and large size are only hydrated.

The following facts will be useful and you should memorize them.

(1) Oxides of nonmetals give rise to an acidic solution in water and the nonmetal behaves as the Lewis acid center in the molecule.

(2) Oxides of metals are basic and will react with protonic Lewis acids (eg. H_2O).

(3) Most chlorides (the entire complex) behave as Lewis acids.

(4) For hydrolysis with cations or anions the extent of hydrolysis decreases as we move down a group.

2. Oxidation – Reduction Reactions

In order to understand oxidation – reduction reactions we must become familiar with the concept of oxidation state, and the idea of species being oxidized and reduced. Using Table 10-4 from the text (reproduced below) let's illustrate what we mean by these terms.

oxid. state of N	compound
+5	HNO_3
+4	NO_2
+3	HNO_2
+2	NO
+1	N_2O
0	N_2

oxid. state of N	compound
-1	NH_2OH
-2	N_2H_4
-3	NH_3

Following the rules outlined in the text we can determine the oxidation state of N in the compounds above. For example in HNO_3 the oxidation state of O is -2 and that of H is +1. The total is then $3 \times -2 + 1 = -5$. Since HNO_3 is a neutral molecule, the N in HNO_3 must have an oxidation state of +5. For N_2H_4, the hydrogens give a sum of +4. Thus the two nitrogens must have a total oxidation number of -4 and each N atom must have an oxidation number of -2. On going from the compound N_2H_4 to HNO_3 the oxidation number of N has changed from <u>-2 to +5</u> and the substance N_2H_4 is said to be oxidized (increase in oxidation state or loss of electrons). On going from the compound HNO_3 to N_2H_4 the oxidation number of N has <u>decreased</u> from +5 to -2 and the substance HNO_3 is said to be reduced (gain of electrons). <u>The reactant that causes an increase in oxidation state is called the oxidizing agent.</u> <u>The reactant that causes a decrease in oxidation state is called the reducing agent.</u> Thus in order to convert a compound at the top to one below it, eg. HNO_3 to NH_3, we require a reducing agent. On going from a compound at the bottom to one above it, eg. NH_3 to HNO_3, we require an oxidizing agent. Fixation of N_2 to form NH_3 requires a reducing agent i.e. in the Haber process

$$N_2 + 3H_2 \longrightarrow 2NH_3$$

148

the reducing agent is H_2 since H changes its oxidation state from 0 in H_2 to +1 in NH_3. All possible terms are summarized in the schematic below

$$\begin{array}{c} \overset{-1e^-}{\boxed{\text{oxidation state } o \rightarrow +1}} \\ N_2 + 3H_2 \longrightarrow 2NH_3 \\ \underset{+3e^-}{\boxed{\text{oxidation state } o \rightarrow -3}} \end{array}$$

H_2 is the reducing agent and becomes oxidized.

N_2 is the oxidizing agent and becomes reduced.

The above example is an equation that is easily balanced by inspection. In many cases it is very difficult to balance oxidation - reduction equations and the method of half - reactions is employed. For example, balance the following equation:

$$Cr_2O_7^{2-} + I^- + H^+ \rightarrow Cr^{3+} + I_2 + H_2O$$

Convert the net ionic equation into two half-reactions, one containing the species reduced and the other the species oxidized.

$$Cr_2O_7^{2-} \longrightarrow Cr^{3+}$$
$$I^- \longrightarrow I_2$$

Balance these half-reactions in terms of atoms. First balance all atoms except H and O.

$$Cr_2O_7^{2-} \longrightarrow 2Cr^{3+}$$
$$2I^- \longrightarrow I_2$$

Determine if the solution is acidic or basic. If acidic, H^+ is added to the side of the equation that has an excess of oxygen atoms and water

149

to the other side to balance the equation in terms of atoms.

$$14H^+ + Cr_2O_7^{2-} \longrightarrow 2\,Cr^{3+} + 7H_2O$$

$$2I^- \longrightarrow I_2$$

Balance the total charges on both sides of each half-reaction by adding electrons to the appropriate side. For the $Cr_2O_7^{2-}$ reaction the charges on the left hand side of the equation total +14 −2 = + 12 while the charges on the right hand side total +6. Therefore we require $6e^-$ on the left hand side.

$$14H^+ + Cr_2O_7^{2-} + 6e^- \longrightarrow 2\,Cr^{3+} + 7H_2O$$

For the I^- reaction we have

$$2I^- \longrightarrow I_2 + 2e^-$$

Now find the least common multiple of the number of electrons in the two half-reactions.

$$14H^+ + Cr_2O_7^{2-} + 6e^- \longrightarrow 2\,Cr^{3+} + 7H_2O$$

$$3 \times [\; 2I^- \longrightarrow I_2 + 2e^- \;]$$

add

$$14H^+ + Cr_2O_7^{2-} + \cancel{6e^-} + 6I^- \longrightarrow 2\,Cr^{3+} + 7H_2O + 3I_2 + \cancel{6e^-}$$

$$14H^+ + Cr_2O_7^{2-} + 6I^- \longrightarrow 2\,Cr^{3+} + 7H_2O + 3I_2$$

*If the solution is basic, OH^- is added to the side of the equation that is deficient in oxygen in order to balance the O atoms. H_2O is added to the other side and the amount of OH^- and H_2O adjusted to balance both H and O atoms.

B. <u>Types of Problems</u>

1. Periodic Trends and Reactions.

2. Oxidation State Determination.

3. Balancing Oxidation-Reduction Reactions.

<u>Examples</u>

1. (i) Predict the formula of the oxides of the following elements and whether you would expect the oxide to be acidic, basic or amphoteric. Write an equation. (a) Sr   (b) Se   (c) B   (d) Al   (e) As.

(ii) Predict the products of the reaction (indicate the intermediate) of:

(a) $SbCl_3 + H_2O$.

(b) $PCl_3 + H_2O$.

(c) $NO_2 + H_2O$.

(d) $BF_3 + (CH_3)_2O$.

(i) (a) Sr is in group 11A and therefore the oxide is SrO. $(Sr^{2+} + O^{2-})$. Since Sr is on the left hand side of the periodic table it has low electronegativity and thus its oxide is basic: (oxide of a metal)

$$SrO + H_2O \longrightarrow Sr^{2+} + 2OH^-.$$

(b) Se is in group V1A and the oxide is SeO_3. (There are many others). Since Se is on the right hand side of the periodic table the oxide is acidic: (oxide of a nonmetal).

$$SeO_3 + H_2O \longrightarrow H_2SeO_4 \ [2H^+ + SeO_2^{2-}].$$

(c) The predicted oxide of B (group 111A) is B_2O_3. Since B is non-

151

metallic, the oxide is acidic:

$$B_2O_3 + 3H_2O \longrightarrow 2B(OH)_3 \ [H_3BO_3].$$

(d)   The predicted oxide of Al (group 111A) is Al_2O_3. Al is right on the dividing line between metals and nonmetals and is <u>amphoteric</u>.

$$Al_2O_3 + 3H_2O \longrightarrow 2Al(OH)_3.$$

(e)   The predicted oxide of As (group VA) is As_2O_5. As is non-metallic and its oxide is acidic.

$$As_2O_5 + 3H_2O \longrightarrow 2H_3AsO_4.$$

(ii) (a) $\underline{SbCl_3} + H_2O$

Since most <u>chlorides</u> (in this case $SbCl_3$) behave as Lewis acids we have the intermediate $[H_2O----SbCl_3]$. This inter-mediate gives the products $SbOCl + 2HCl$.

(b) $\underline{PCl_3} + H_2O$

The intermediate is $[H_2O---PCl_3]$. In this case the products are $H_3PO_3 + 3HCl$.

(c) $\underline{NO_2} + H_2O$

NO_2 is an oxide of a very electronegative element and behaves as a Lewis acid giving the interm $[H_2O----NO_2]$ which gives the products HNO_3 and $NO(g)$.

(d) $\underline{BF_3} + (CH_3)_2O$

Here BF_3 is the Lewis acid and $(CH_3)_2O$ the Lewis base.

$$[\ (CH_3)_2O----BF_3 \ ] \longrightarrow (CH_3)_2 \ O-BF_3$$

2.   Calculate the oxidation state of the underlined atom for (a)-(d) on the next page.

152

(a) $Na_2\underline{C}_2O_4$ (b) $H\underline{As}O_4{}^{2-}$ (c) $(NH_4)_2\underline{S}_2O_8$ (d) $\underline{C}S_2$.

(a) $Na_2C_2O_4$

$$\left.\begin{array}{l} 2Na = +2 \\ 4\ O = -8 \\ \therefore\ 2C = +6 \end{array}\right\} \quad \text{Total} = 0$$

$$\therefore\ C = +3$$

(b) $HAsO_4{}^{2-}$

$$\left.\begin{array}{l} 1H = +1 \\ 4\ O = -8 \\ \therefore\ As = +5 \end{array}\right\} \quad \text{Total} = -2$$

(c) $(NH_4)_2S_2O_8$

$$\left.\begin{array}{l} 2NH_4 = +2 \\ 8\ O = -16 \\ \therefore\ 2S = +14 \end{array}\right\} \quad \text{Total} = 0$$

$$\therefore\ S = +7$$

(d) CS_2

$$2S = -4$$

$$\therefore\ C = +4$$

3. (i) Balance the following equation.

$$AsO_3{}^{3-} + MnO_4{}^- + H^+ \longrightarrow AsO_4{}^{3-} + Mn^{2+} + H_2O.$$

(ii) Question 15. at the end of the chapter. When a solution of hydrazine is mixed with a solution of Br_2, N_2 gas is evolved and the solution is decolorized. Write a balanced equation.

(i) Write the two half-reactions

$$\left.\begin{array}{l} AsO_3{}^{3-} \longrightarrow AsO_4{}^{3-} \\ MnO_4{}^- \longrightarrow Mn^{2+} \end{array}\right\} \quad \begin{array}{l}\text{All atoms except O} \\ \text{are balanced.}\end{array}$$

Add H^+ to the side of the equation that has an excess of oxygen atoms and water to the other side to balance the equation in terms of atoms.

$$H_2O + AsO_3^{3-} \longrightarrow AsO_4^{3-} + 2H^+$$

$$8H^+ + MnO_4^- \longrightarrow Mn^{2+} + 4H_2O$$

Balance the total charges on both sides of each half-reaction by adding electrons to the appropriate side. For the AsO_3^{3-} reaction the total charge on the left hand side $= -3$ while on the right hand side the total charge is -1. Therefore add $2e^-$ to the right hand side:

$$H_2O + AsO_3^{3-} \longrightarrow AsO_4^{3-} + 2H^+ + 2e^-$$

For the MnO_4^- reaction, the charge on the left hand side $= +7$ and on the right hand side $= +2$. Therefore:

$$8H^+ + MnO_4^- + 5e^- \longrightarrow Mn^{2+} + 4H_2O$$

Find the least common multiple of the number of electrons in the two half-reactions:

$$5 \times [H_2O + AsO_3^{3-} \longrightarrow AsO_4^{3-} + 2H^+ + 2e^-]$$

$$2 \times [8H^+ + MnO_4^- + 5e^- \longrightarrow Mn^{2+} + 4H_2O]$$

add

$$5H_2O + 5AsO_3^{3-} + 16H^+ + 2MnO_4^- \longrightarrow 5AsO_4^{3-} + 10H^+ + 2Mn^{2+} + 8H_2O$$

$$5AsO_3^{3-} + 6H^+ + 2MnO_4^- \longrightarrow 5AsO_4^{3-} + 2Mn^{2+} + 3H_2O$$

(ii)   From table 10-5 we know that

$$N_2H_4 \longrightarrow N_2$$

$$\text{and} \quad Br_2 \longrightarrow Br^-$$

154

The half reactions are

$$N_2H_4 \longrightarrow N_2 + 4H^+ + 4e^-$$

$$2e^- + Br_2 \longrightarrow 2Br^-$$

Multiply the Br_2 reaction x 2:

$$4e^- + 2Br_2 \longrightarrow 4\ Br^-$$

$$N_2H_4 \longrightarrow N_2 + 4H^+ + 4e^-$$

add:

$$N_2H_4 + 2Br_2 \longrightarrow N_2 + 4Br^- + 4H^+$$

C.   <u>Test Yourself</u>

1. (i)   Predict the formulas of the sulfides of Rb, Co(II), and Ca.

   (ii) Predict the formulas of the oxides of Se, S, Cl, and Rb and
         whether you would expect the oxide to react with water to give
         an acidic or basic solution. Write an equation.

   (iii) Predict the products of the reaction of

         (a) CsCl and H_2O

         (b) $Cr(ClO_4)_3$ and H_2O

         (c) $BeCl_2$ and H_2O

   (iv) Predict the products of the reaction (indicate the intermediate
         in (a), (b), (e)) of:

         (a) $LiH + H_2O$

         (b) $BeH_2 + NH_3$

         (c) $N_2O_5 + H_2O$

         (d) $NCl_3 + H_2O$

         (e) $BiCl_3 + H_2O$

2.   Calculate the oxidation state of the underlined atom.

        (a) $H_2\underline{S}O_4$

        (b) $H_2\underline{S}O_3$

        (c) $K\underline{I}O_4$

3.   Balance the following equations:

        (a) $H_2O_2 + MnO_4^- + H^+ \longrightarrow O_2 + Mn^{2+} + H_2O$

        (b) $Na_2S_2O_3 + KMnO_4 + H_2O \longrightarrow Na_2SO_4 + K_2SO_4 + MnO_2 + KOH$

156

(c) $CN^- + IO_3^- \longrightarrow I^- + CNO^-$ (basic solution).

4.   An acid solution of $Cr_2O_7^{2-}$ and NO_2^- are mixed and the solution changes color from orange to a pale violet. Write a balanced chemical equation for this reaction.

5. (a) Calculate the oxidation state of N in each of the following species:

N_2, N_2H_2, N_2H_4, NH_3.

(b)   The products of biological nitrogen (N_2) fixation are the expected NH_3 as well as H_2. Write a balanced half-reaction for this process.

157

11

Enthalpy Changes Accompanying Chemical Reactions

A. Points of Importance

1.   Memorize the equation $\Delta H = \Delta E + (\Delta n)RT$ where Δn is the
final number of moles of gas in the product minus the original number
of moles of gas in the reactants.

($RT = 593$ cal mole^{-1} at 25°C, $R = 1.987$ cal mole^{-1}deg^{-1}).

2.   The enthalpy of formation, ΔH_f°, of a compound is defined as the
enthalpy change that occurs when one mole of the compound is formed
from the elements in their stable forms at 25°C and 1 atmosphere
pressure. The enthalpy of formation of any element in its stable form
at 25°C and 1 atmosphere pressure is zero. It is important to become
proficient at writing equations corresponding to a ΔH_f° reaction. For
example, the equation for the enthalpy of formation of ethanol, $C_2H_5OH(\ell)$
is $2C$ (graphite) $+ 3H_2(g) + \frac{1}{2}O_2(g) \longrightarrow C_2H_5OH(\ell)$.

There are two important points to note about this equation:

(a) the physical state of each species must be specified i.e. s, ℓ, g
     means solid, liquid, gas.

(b) the reaction is written for one mole of product and therefore
     fractional coefficients for reactants are acceptable.

Just as you need to know what equation corresponds to ΔH_f° there are a
few other reactions which are met very often that you should know. Some
of these are summarized below using examples:

$$C(s) + O_2(g) \longrightarrow CO_2(g) \quad \Delta H^\circ_{rxn} = \Delta H_f^\circ (CO_2)$$

$$C_2H_5OH(\ell) + 3O_2(g) \longrightarrow 2CO_2(g) + 3H_2O(g)$$

$$\Delta H^\circ_{rxn} = \Delta H^\circ_{combustion}$$

$$H_2O(\ell) \longrightarrow H_2O(g) \quad \Delta H^\circ_{rxn} = \Delta H^\circ_{vaporization}$$

$$NO(g) + \tfrac{1}{2}O_2(g) \longrightarrow NO_2(g) \quad \Delta H^\circ_{rxn} = \Delta H^\circ_{comubstion}$$

$$H_2O(s) \longrightarrow H_2O(g) \quad \Delta H^\circ_{rxn} = \Delta H^\circ_{sublimation}$$

For the formation of $CO_2(g)$, the equation could also be referred to as
a combustion reaction since $C(s)$ is burned in $O_2(g)$ just as in the
combustion of $C_2H_5OH(\ell)$.

   Remember that ΔH whether labelled ΔH_f°, $\Delta H_{combustion}$, $\Delta H_{vaporization}$,
or $\Delta H_{sublimation}$ always refers to a particular reaction and you must be
able to write an equation for the particular process. The subscripts
are only labels but know what they mean.

   The enthalpy change, ΔH°_{rxn}, for any reaction is the sum of the
enthalpies of formation of the products minus the sum of the enthalpies

of formation of the reactants:

$$\Delta H^{\circ}_{rxn} = \Sigma n_p \Delta H^{\circ}_f \text{ (products)} - \Sigma n_r \Delta H^{\circ}_f \text{ (reactants)}$$

This equation can be used to solve most thermochemistry problems and its use will be illustrated in section B.

An additional method of estimating ΔH°_{rxn} is also available using the data in Table 11-3. The data here are given as bond dissociation energies and the values are all positive indicating that energy is required to break bonds. Thus if a bond is being formed the same amount of energy must be liberated. Therefore the numbers in Table 11-3 multiplied by −1 can be taken to represent the heat of bond formation. In this case the equation to be used is

$$\Delta H^{\circ}_{rxn} = \Sigma \Delta H^{\circ}_{bf} \text{ (product bonds)} - \Sigma \Delta H^{\circ}_{bf} \text{ (reactant bonds)}.$$

To estimate the heat of formation of $H_2O(g)$ first write the equation:

$$H-H(g) \quad + \quad \tfrac{1}{2} \; O=O(g) \longrightarrow \quad _H \diagdown^O\diagdown_H \quad (g)$$

$$\Delta H^{\circ}_{rxn} = \Delta H^{\circ}_f \; (H_2O(g)) = [2(\Delta H^{\circ}_{bf}(O-H))] - [\tfrac{1}{2}(\Delta H^{\circ}_{bf}(O=O)) + \Delta H^{\circ}_{bf}(H-H)]$$

$$= [2(-110)]-[\tfrac{1}{2}(-118) + (-104)]$$

$$= -220 + 59 + 104$$

$$= -57 \text{ kcal mole}^{-1}.$$

To estimate ΔH°_{rxn} using bond energies all reactants and products must be in the gas phase. A thermodynamic cycle for the reaction just

160

discussed can also be constructed

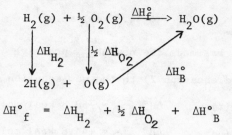

$$\Delta H^{\circ}{}_{f} = \Delta H_{H_2} + \tfrac{1}{2} \Delta H_{O_2} + \Delta H^{\circ}{}_{B}$$

In this case ΔH_{H_2} and $\tfrac{1}{2} \Delta H_{O_2}$ are both positive since they represent

the dissociation of H_2 and O_2. But ΔH°_{B} is negative since it represents

<u>formation</u> of two O-H bonds i.e. $\Delta H^{\circ}_{f} = 104 + \tfrac{1}{2} \times 118 + 2\,(-110) = -57$

kcal mole^{-1}. Thus one can see the two methods are equivalent. Use the

easiest method for the particular problem.

3. <u>Hess's Law and Enthalpy – Energy Cycles.</u>

     Since ΔH is a state function we can add and subtract equations

(and ΔH's) to obtain a desired equation (and ΔH). For example, say

we are given the following equations:

1) $H_2(g) + O_2(g) \longrightarrow H_2O_2(\ell)$ $\Delta H^{\circ}_{rxn}\,①\, = -44.8$ kcal mole^{-1}

2) $H_2(g) + \tfrac{1}{2}O_2(g) \longrightarrow H_2O(\ell)$ $\Delta H^{\circ}_{rxn}\,②\, = -68.3$ kcal mole^{-1}

and we want to know H°_{rxn} for the decomposition of hydrogen peroxide,

$H_2O_2(\ell)$:

3) $H_2O_2(\ell) \longrightarrow \tfrac{1}{2}O_2(g) + H_2O(\ell)$. Hess's Law is applied in the

    following way.

     Since reaction 3) has $H_2O_2(\ell)$ on the left hand side we need to

reverse reaction 1) to obtain:

4) $H_2O_2(\ell) \longrightarrow H_2(g) + O_2(g)$ $\Delta H^\circ_{rxn} \text{④} = -\Delta H^\circ_{rxn} \text{①}$

$$= + 44.8 \text{ kcal mole}^{-1}.$$

Note that reversing reaction 1) has made the new ΔH°_{rxn} positive (does this make sense?). Now add equation 2) to the new equation 4):

$H_2(g) + \frac{1}{2} O_2(g) \longrightarrow H_2O(\ell)$ $\Delta H^\circ_{rxn} \text{②} = -68.3 \text{ kcal mole}^{-1}$

$H_2O_2(\ell) \longrightarrow H_2(g) + O_2(g)$ $\Delta H^\circ_{rxn} \text{④} = +44.8 \text{ kcal mole}^{-1}$

add:

$H_2O_2(\ell) \longrightarrow \frac{1}{2}O_2(g) + H_2O(\ell)$ $\Delta H^\circ_{rxn} = -23.5 \text{ kcal mole}^{-1}$

We have obtained the ΔH° of the required equation by manipulating known equations. The use of Hess's Law is summarized below:

1) ΔH°_{rxn} depends on the number of moles of substance reacted or produced. If we had decomposed 2 moles of $H_2O_2(\ell)$ the ΔH°_{rxn} would have been 2 x -23.5 or -47.0 kcal mole^{-1}.

2) When an equation is reversed, the sign of ΔH°_{rxn} is also changed. Equation 1) was reversed and the ΔH°_{rxn} changed from negative to positive.

3) ΔH°_{rxn} for a particular reaction which is made up of two different reactions is simply the sum of the ΔH°_{rxn}'s for the two different reactions. Addition of equations 2) and 4) gave the required equation and the ΔH° required was obtained from the sum of $\Delta H^\circ_{rxn} \text{②} + \Delta H^\circ_{rxn} \text{④}$

Hess's Law can also be used to construct energy or enthalpy

162

cycles and these are fully discussed in sections 11-4 and 11-5.

Finally note that the SI unit of energy is the joule where
1 calorie = 4.184 joule or 1 kcal = 4.184 kJ.

B.   Types of Problems

    * Problems are worked both in kcal and kJ (1kcal = 4.184 kJ).

1. Application of the equation $\Delta H = \Delta E + (\Delta n)RT$.

2. Write equations representing ΔH_f° or any ΔH_{rxn}°.

3. Application of the equation

$$\boxed{\Delta H_{rxn}^\circ = \Sigma n_p \Delta H_f^\circ \text{ (products)} - \Sigma n_r \Delta H_f^\circ \text{ (reactants)}}$$

4. Application of Hess's Law.

    For 3. and 4. above one is usually faced with two types

    of problems:

    (a) given ΔH_f°, calculate ΔH_{rxn}°

    (b) given ΔH_{rxn}°, calculate ΔH_f°.

5. Heats of bond formation and bond dissociation.

Examples

1.(i) Calculate ΔE° for the reaction

$$N_2H_4(\ell) + O_2(g) \longrightarrow N_2(g) + 2H_2O(g)$$

given $\Delta H_f^\circ (N_2H_4(\ell)) = 12.1$ kcal mole^{-1} and

    $\Delta H_f^\circ (H_2O(g)) = -57.8$ kcal mole^{-1}.

(ii) For the reaction

$$\underset{H}{\overset{H}{{}}}N-N\overset{CH_3}{\underset{CH_3}{{}}} (\ell) + 2N_2O_4(\ell) \longrightarrow 3N_2(g) + 2CO_2(g) + 4H_2O(g)$$

$\Delta E^\circ = -424.137$ kcal mole^{-1}. (a) Calculate ΔH° for the reaction.

(b) Calculate the ΔH_f° for $NH_2N(CH_3)_2(\ell)$

if $\Delta H_f^\circ (N_2O_4(\ell)) = -4.66$ kcal mole^{-1}

$\Delta H_f^\circ (CO_2(g)) = -94.1$ kcal mole^{-1}

$\Delta H_f^\circ (H_2O(g)) = -57.8$ kcal mole^{-1}.

(c) Construct an energy level diagram to illustrate ΔH°, ΔE°, and W.

(i) In order to obtain ΔE° we require ΔH° and then need to use the equation $\Delta H^\circ = \Delta E^\circ + (\Delta n)RT$.

$$N_2H_4(\ell) + O_2(g) \longrightarrow N_2(g) + 2H_2O(g)$$

$$\Delta H_{rxn}^\circ = \Sigma n_p \Delta H_f^\circ \text{ (products)} - \Sigma n_r \Delta H_f^\circ \text{ (reactants)}$$

$$= 2(-57.8) - (12.1)$$

$$= -115.6 - 12.1$$

$$= -127.7 \text{ kcal mole}^{-1}$$

$\Delta n = +2$ [moles gaseous products - moles gaseous reactants]

$$\therefore \Delta E_{rxn}^\circ = \Delta H_{rxn}^\circ - (\Delta n)RT.$$

$$= -127,700 - (2)1.987 \times 298 \text{ cal mole}^{-1}$$

$$= -127,700 - (2)593$$

$$= -127,700 - 1,186$$

$$= -128,900 \text{ cal mole}^{-1}$$

$$= -128.9 \text{ kcal mole}^{-1}.$$

(ii) (a) For this reaction we can use $\Delta H^\circ = \Delta E^\circ + (\Delta n)RT$ immediately.

$$\Delta n = +9$$

$$\therefore \Delta H_{rxn}^\circ = -424,137 + (9)1.987 \times 298 \text{ cal mole}^{-1}$$

$$= -424,137 + 5,337$$

$$= -418,800 \text{ cal mole}^{-1}$$

$$= -418.800 \text{ kcal mole}^{-1}.$$

(b) Since we have ΔH°_{rxn} we can use the equation

$\Delta H^\circ_{rxn} = \Sigma n_p \Delta H^\circ_f$ (products) $- \Sigma n_r \Delta H^\circ_f$ (reactants) to calculate

ΔH°_f ($NH_2N(CH_3)_2(\ell)$).

$\Delta H^\circ_{rxn} = [2(-94.1) + 4(-57.8)] - [2(-4.66) + \Delta H^\circ_f(NH_2N(CH_3)_2(\ell))]$

$-418.800 = -188.2 - 231.2 + 9.32 - \Delta H^\circ_f(NH_2N(CH_3)_2(\ell))$

$\Delta H^\circ_f(NH_2N(CH_3)_2(\ell)) = -188.2 - 231.2 + 9.32 + 418.800$

$= + 8.7$ kcal mole^{-1}

(c)

$NH_2N(CH_3)_2(\ell) + 2N_2O_4(\ell)$ _____ 0

$-\Delta H^\circ$

ΔE°

$3N_2(g) + 2CO_2(g) + 4H_2O(g)$_____ $-418,800$ cal mole^{-1}

$3N_2(g) + 2CO_2(g) + 4H_2O(g)$___$-\Delta nRT$ $-424,137$ cal mole^{-1}

If $\underline{\Delta n \text{ is positive}}$, the term ΔnRT will make a negative contribution to ΔE° and $\underline{\text{work is done by the system}}$. In sample problem 2, $\underline{\Delta n \text{ is negative}}$, the term ΔnRT makes a positive contribution to ΔE° and $\underline{\text{work is done on the}}$ $\underline{\text{system}}$. In this particular example $\underline{\Delta n = + 9}$ and the atmosphere is being moved back to accomodate more gas at constant pressure and therefore $\underline{\text{work is done by the system}}$.

Remember for a constant volume process

$q_v = \Delta E$

and for a constant pressure process

$q_p = \Delta H$.

2.(i) Write the equation which represents the heat of formation of the following.

(a) $H_2N-N(CH_3)_2(\ell)$

(b) $CaCO_3(s)$

(c) $N_2H_4(\ell)$

(d) $SO_3(g)$

(e) $As_2O_3(s)$

(ii) Write the equation which represents the heat of combustion of the following at 25°C and 1 atm.

(a) $C_6H_{12}O_6(s)$

(b) $C_6H_6(\ell)$

(c) $NO(g)$

(d) $H_2(g)$

(i) (a) $N_2(g) + 4H_2(g) + 2C(s) \longrightarrow H_2N-N(CH_3)_2(\ell)$

(b) $Ca(s) + C(s) + \frac{3}{2}O_2(g) \longrightarrow CaCO_3(s)$

(c) $N_2(g) + 2H_2(g) \longrightarrow N_2H_4(\ell)$

(d) $S(s) + \frac{3}{2}O_2(g) \longrightarrow SO_3(g)$

(e) $2As(s) + \frac{3}{2}O_2(g) \longrightarrow As_2O_3(s)$

(ii) (a) $C_6H_{12}O_6(s) + 6O_2(g) \longrightarrow 6CO_2(g) + 6H_2O(\ell)$

(b) $C_6H_6(\ell) + \frac{15}{2}O_2(g) \longrightarrow 6CO_2(g) + 3H_2O(g)$

(c) $NO(g) + \frac{1}{2}O_2(g) \longrightarrow NO_2(g)$

(d) $H_2(g) + \frac{1}{2}O_2(g) \longrightarrow H_2O(g)$

Note that this reaction also represents the heat of formation of $H_2O(g)$.

3.(i) Calculate ΔH°_{rxn} for the following reactions.

(a) $C_2H_2(g) + \frac{5}{2}O_2(g) \longrightarrow 2CO_2(g) + H_2O(g)$

(b) $2SO_2(g) + O_2(g) \longrightarrow 2SO_3(g)$

(c) $Mg(s) + \frac{1}{2}O_2(g) \longrightarrow MgO(s)$

Given: $\Delta H^\circ_f(C_2H_2(g)) = 54.2$ kcal, $\Delta H^\circ_f(CO_2(g)) = -94.1$ kcal,

$\Delta H^\circ_f(H_2O(g)) = -57.8$ kcal, $\Delta H^\circ_f(SO_2(g)) = -70.9$ kcal,

$\Delta H^\circ_f(SO_3(g)) = -94.6$ kcal, $\Delta H^\circ_f(MgO(s)) = -143.8$ kcal.

(ii) The enthalpy of formation of n-pentane, $C_5H_{12}(\ell)$, is -41.4 kcal mole^{-1}.

Calculate the enthalpy change for combustion of 5.00 g of n-pentane

to $CO_2(g)$ and $H_2O(\ell)$. Given: $\Delta H^\circ_f(CO_2(g)) = -94.1$ kcal, $\Delta H^\circ_f(H_2O(\ell)) =$

-68.3 kcal.

(iii) The heat of combustion of nitromethane, $CH_3NO_2(\ell)$, at 25° and 1

atmosphere is -152.6 kcal mole^{-1}. Calculate the heat of formation

of $CH_3NO_2(\ell)$. Given: $\Delta H^\circ_f(CO_2(g)) = -94.1$ kcal, $\Delta H^\circ_f(H_2O(g)) = -57.8$ kcal.

(iv) Calculate the heat of formation of $N_2H_4(\ell)$ given the heat of

combustion of $N_2H_4(\ell)$ of -149.0 kcal mole^{-1}. (The products are

$N_2(g)$ and $H_2O(\ell)$). $\Delta H^\circ_f(H_2O(\ell)) = -68.3$ kcal.

(i) For (a), (b), and (c) the equation

$$\Delta H^\circ_{rxn} = \Sigma n_p \Delta H^\circ_f \text{ (products)} - \Sigma n_r \Delta H^\circ_f \text{ (reactants) is used.}$$

(a) $\Delta H^\circ_{rxn} = [2(-94.1) + (-57.8)] - [54.2]$

$= -300.2$ kcal mole^{-1}.

(b) $\Delta H^\circ_{rxn} = [2(-94.6)] - [2(-70.9)]$

$= -47.4$ kcal mole^{-1}.

(c) $\Delta H^\circ_{rxn} = [(-143.8)] - [0]$

$= -143.8$ kcal mole^{-1}.

This is the heat of formation for MgO(s).

(ii) First write a balanced equation for combustion of one mole of n-pentane.

$$C_5H_{12}(\ell) + 8\ O_2(g) \longrightarrow 5CO_2(g) + 6H_2O(\ell).$$

Now calculate ΔH°_{rxn} for combustion of one mole of n-pentane.

$$\Delta H^\circ_{rxn} = [5(-94.1) + 6(-68.3)]-[(-41.4)]$$

$$= -838.9 \text{ kcal mole}^{-1}$$

$$\text{number of moles } C_5H_{12} = \frac{5.00g}{72.0\frac{g}{mole}} = 0.0694.$$

Therefore 0.0694 moles x $-838.9\ \dfrac{kcal}{mole} = -58.2$ kcal.

This is the enthalpy change for combustion of 5.00 g $C_5H_{12}(\ell)$.

(iii) Again write the equation.

$$CH_3NO_2(\ell) + \frac{3}{4}\ O_2(g) \longrightarrow CO_2(g) + \frac{3}{2}\ H_2O(g) + \frac{1}{2}\ N_2(g)$$

In this case we are given ΔH°_{rxn} and need only to solve for $\Delta H^\circ_f(CH_3NO_2(\ell))$.

$$\Delta H^\circ_{rxn} = [\Delta H^\circ_f((CO_2)(g)) + \frac{3}{2}\ \Delta H^\circ_f((H_2O(g))]-[\Delta H^\circ_f(CH_3NO_2(\ell))]$$

$$-152.6 = [(-94.1) + \frac{3}{2}(-57.8)] - \Delta H^\circ_f(CH_3NO_2(\ell))$$

$$\Delta H^\circ_f(CH_3NO_2(\ell)) = 152.6 + (-94.1) + \frac{3}{2}\ (-57.8)$$

$$= 152.6 - 180.8$$

$$= -28.2 \text{ kcal mole}^{-1}.$$

(iv) Write the equation for combustion.

$$N_2H_4(\ell) + O_2(g) \longrightarrow N_2(g) + 2H_2O(\ell)$$

$$-149.0 = [2(-68.3)] - [\Delta H_f^\circ(N_2H_4(\ell))]$$

$$\Delta H_f^\circ(N_2H_4(\ell)) = 149.0 - 136.6$$

$$= +12.4 \text{ kcal mole}^{-1}.$$

4.(i)   Some "beetles" defend themselves by spraying hot <u>quinone</u> at their attackers. Calculate "ΔH_{beetle}°" for

$$C_6H_4(OH)_2(\ell) + H_2O_2(\ell) \longrightarrow C_6H_4O_2(\ell) + 2H_2O(\ell)$$

hydroquinone   <u>quinone</u>

given:

$$C_6H_4(OH)_2(\ell) \longrightarrow C_6H_4O_2(\ell) + H_2(g) \qquad \Delta H_1^\circ = +177.4 \text{ kJ}$$

$$O_2(g) + H_2(g) \longrightarrow H_2O_2(\ell) \qquad \Delta H_2^\circ = -187.4 \text{ kJ}$$

$$H_2(g) + \tfrac{1}{2} O_2(g) \longrightarrow H_2O(\ell) \qquad \Delta H_3^\circ = -285.8 \text{ kJ}$$

(ii)   Construct a Born-Haber cycle for the reaction

$$Li(s) + \tfrac{1}{2} Cl_2(g) \longrightarrow Li^+Cl^-(s).$$

Given: ΔH_D of $Cl_2 = +58 \text{ kcal mole}^{-1}$, IE of Li = $+124$ kcal mole^{-1}, ΔH_S Li = $+38.6 \text{ kcal mole}^{-1}$, EA of chlorine = $-83 \text{ kcal mole}^{-1}$, and $\Delta H_f^\circ = -97.6$ kcal mole^{-1}. Calculate the lattice energy, ΔH_L.

(i) Reverse ΔH_2° and add to $2\Delta H_3^\circ$:

$$H_2O_2(\ell) \longrightarrow O_2(g) + H_2(g) \qquad \Delta H_{-2}^\circ = +187.4 \text{ kJ}$$

$$2H_2(g) + O_2(g) \longrightarrow 2H_2O(\ell) \qquad 2\Delta H_3^\circ = -571.6 \text{ kJ}$$

add:
_____

$$H_2O_2(\ell) + H_2(g) \longrightarrow 2H_2O(\ell) \qquad \Delta H_4^\circ = \Delta H_{-2}^\circ + 2\Delta H_3^\circ$$

$$= -384.2 \text{ kJ}$$

Now add $\Delta H_4^\circ + \Delta H_1^\circ$:

$$H_2O_2(\ell) + H_2(g) \longrightarrow 2H_2O(\ell) \qquad \Delta H_4^\circ = -384.2 \text{ kJ}$$

$$C_6H_4(OH)_2(\ell) \longrightarrow C_6H_4O_2(\ell) + H_2(g) \qquad \Delta H_1^\circ = +177.4 \text{ kJ}$$

add:

$$C_6H_4(OH)_2(\ell) + H_2O_2(\ell) \longrightarrow C_6H_4O_2(\ell) + 2H_2O(\ell)$$

$$\Delta H_{beetle}^\circ = \Delta H_4^\circ + \Delta H_1^\circ$$

$$= -206.8 \text{ kJ}$$

(ii)

$$\Delta H_f^\circ = \Delta H_s + IE + \tfrac{1}{2}\Delta H_D + EA + \Delta H_L$$

$$-97.6 = 38.6 + 124 + \tfrac{1}{2} \times 58 - 83 + \Delta H_L$$

$$\Delta H_L = -206.2 \text{ kcal mole}^{-1}.$$

5.(i)   Question 14 at the end of chapter 11. Using the heat of formation of ethane, H_3C-CH_3 in Table 11-2, the C—H bond energy in Table 11-3, the heat of atomization of graphite of 172,000 cal mole^{-1}, and the heat of dissociation of H_2 of 104,180 cal mole^{-1}, calculate the C—C bond dissociation energy.

(ii)   Question 15 at the end of chapter 11. Estimate the heat of formation of propane ($CH_3CH_2CH_3$) using the data in Table 11-3.

171

The ΔH_A of atomization for carbon is + 172 kcal mole^{-1}.

(iii) Question 17 at the end of chapter 11. Using bond energy data

(Table 11-3) and a heat of sublimation for I_2 of 15.0 kcal mole^{-1},

calculate the heat of formation of ICl.

(i)   Construct a thermodynamic cycle:

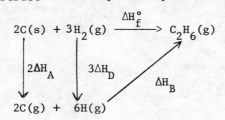

$\Delta H_f^\circ = 2\Delta H_A + 3\Delta H_D + \Delta H_B$ where ΔH_A = heat of atomization

$-20.24 = 2(172) + 3(104.18) + \Delta H_B$ ΔH_D = heat of dissociation

$\Delta H_B = -20.24 - 344 - 312.54$ ΔH_B = heat of bonds forma-
                                                     mation.

$\Delta H_B = - 677$ kcal

This represents the energy given off in the formation of 6 H-C bonds

and 1 C-C bond i.e. from the structure of ethane

$$
\begin{array}{c}
\ \ \text{H} \ \text{H} \\
\ \ | \ \ \ | \\
\text{H--C--C--H.} \\
\ \ | \ \ \ | \\
\ \ \text{H} \ \text{H}
\end{array}
$$

From Table 11-3 we can obtain the value for 6 C-H bonds <u>i.e.</u> 6 x -99 =

-594 kcal. The amount left over must be the heat of bond formation

for a C-C bond.

$$-677 - (-594) = \underline{-83 \text{ kcal.}}$$

Note that +83 kcal represents the amount of energy required to break

one C-C bond or the heat of dissociation.

(ii)   Construct a thermodynamic cycle:

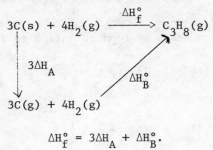

$$\Delta H_f^{\circ} = 3\Delta H_A + \Delta H_B^{\circ}.$$

We can obtain ΔH_B° from Table 11-3 as in the example in the text or simply by using the <u>heats of bond formation</u> instead of dissociation.

$$\Delta H_B^{\circ} = \Sigma \; \Delta H_{bf}^{\circ} \text{ (product bonds)} - \Sigma \; \Delta H_{bf}^{\circ} \text{ (reactant bonds)}$$

For the equation:

$$3C(g) + 4(H-H)(g) \xrightarrow{\Delta H^{\circ}_B} \underset{\substack{\text{H H H}}}{\overset{\substack{\text{H H H}}}{\text{H-C-C-C-H}}}(g)$$

we have:

$$\Delta H_B^{\circ} = [8\Delta H^{\circ}(C-H) + 2\Delta H^{\circ}(C-C)] - [4\Delta H^{\circ}(H-H)]$$

$$= [8(-99) + 2(-80)] - [4-(-104)]$$

$$= -792 \; -160 + 416$$

$$= -536 \text{ kcal.}$$

$$\therefore \quad \Delta H_f^{\circ} = 3(+172) + (-536)$$

$$= -20 \text{ kcal.}$$

(iii) Construct a thermodynamic cycle:

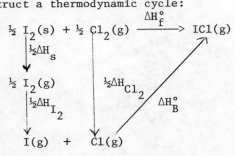

173

$$\Delta H_f^\circ = \tfrac{1}{2} \Delta H_s + \tfrac{1}{2} \Delta H_{I_2} + \tfrac{1}{2} \Delta H_{Cl_2} + \Delta H_B^\circ.$$

Again we obtain ΔH_B° from heats of bond formation in Table 11-3:

$$\Delta H_B^\circ = [\Delta H^\circ \ (I-Cl)]$$

$$= -51 \ \text{kcal}$$

$$\Delta H_f^\circ = \tfrac{1}{2}(15.0) + \tfrac{1}{2}(36) + \tfrac{1}{2}(58) + (-51)$$

$$= 7.5 + 18 + 29 - 51$$

$$= + 4 \ \text{kcal}.$$

C. Test Yourself

1. Calculate $\Delta E°$ for the reaction

$$N_2(g) + O_2(g) \longrightarrow 2NO(g)$$

given $\Delta H_f°(NO(g)) = +21.60$ kcal.

2. Write the reaction which represents the

   (a) combustion of "iso-octane", $C_8H_{18}(\ell)$, to $CO_2(g)$ and $H_2O(g)$.

   (b) heat of formation of $PbCO_3(s)$.

   (c) combustion of coal, C(graphite).

3. Calculate $\Delta H°_{rxn}$ for the reaction which occurs in a catalytic converter on automobiles:

$$2NO(g) + 2CO(g) \longrightarrow N_2(g) + 2CO_2(g)$$

Given:

$$\Delta H_f°(NO(g)) = +21.6 \text{ kcal.}$$

$$\Delta H_f°(CO(g)) = -26.4 \text{ kcal.}$$

$$\Delta H_f°(CO_2(g)) = -94.1 \text{ kcal.}$$

4. Given $4NH_3(g) + 5O_2(g) \longrightarrow 4NO(g) + 6H_2O(\ell)$ $\Delta H°_{rxn} = -1.169 \times 10^3$ kJ

   $4NH_3(g) + 3O_2(g) \longrightarrow 2N_2(g) + 6H_2O(\ell)$ $\Delta H°_{rxn} = -1.530 \times 10^3$ kJ

   Calculate the heat of formation of NO(g).

5. Calculate the heat of formation of $Pb_3O_4(s)$ from the following:

$$Pb_3O_4(s) + 4CO(g) \longrightarrow 3Pb(s) + 4CO_2(g) \quad \Delta H°_{rxn} = -397.3 \text{ kJ}$$

$$C(s) + \tfrac{1}{2} O_{2(g)} \longrightarrow CO(g) \qquad\qquad \Delta H°_{rxn} = -110.5 \text{ kJ}$$

$$C(s) + O_2(g) \longrightarrow CO_2(g) \qquad\qquad \Delta H°_{rxn} = -393.5 \text{ kJ}$$

6. Calculate the heat of formation of the H—Br bond from the following data:

$$H_2(g) + Br_2(g) \longrightarrow 2HBr(g) \qquad \Delta H° = -103 \text{ kJ}$$

$$H_2(g) \longrightarrow 2H(g) \qquad \Delta H° = 436 \text{ kJ}$$

$$Br_2(g) \longrightarrow 2Br(g) \qquad \Delta H° = 193 \text{ kJ}$$

7. Using Table 11-3 calculate the $\Delta H°_{rxn}$ (heat of combustion) for combustion of "iso-octane" $C_8H_{18}(\ell) + 12\frac{1}{2} O_2(g) \longrightarrow 8CO_2(g) + 9H_2O(g)$ given the heat of vaporization of $C_8H_{18}(\ell)$ of 9.22 kcal mole^{-1}. The structure of "iso-octane" is

8. Calculate the lattice energy, ΔH_L, for Li^+I^- given ΔH_D of I_2 = + 36 kcal mole^{-1}, IE of Li = + 124 kcal mole^{-1}, ΔH_s Li = + 38.6 kcal mole^{-1}, EA of iodine = -71 kcal mole^{-1}, and $\Delta H°_f$ = -64.6 kcal mole^{-1}.

9. Calculate the heat of formation of $Fe_2O_3(s)$ given $\Delta H°_{rxn}$ = - 791 kcal mole^{-1} for $4FeS_2(s) + 11O_2(g) \longrightarrow 2Fe_2O_3(s) + 8SO_2(g)$, $\Delta H°_f$ $(SO_2(g))$ = - 71.0 kcal mole^{-1}, and $\Delta H°_f(FeS_2(s))$ = - 42.5 kcal mole^{-1}.

12

Quantitative Treatment of Acid-Base Equilibria

A. Points of Importance

1. pH and Water.

The pH of a solution is defined as the negative logarithm of the hydrogen ion concentration:

$$\boxed{pH = - \log [H^+]} \qquad \text{(memorize)}.$$

This provides a convenient measure of H^+ ion in any solution on a scale of 1-14 at 25°C. This is based on the ion product of water

$$[H^+][OH^-] = 1.0 \times 10^{-14} \text{ at } 25°C.$$

Other quantities are also defined using the "p" notation ("p" stands for $- \log$):

$$pKa = - \log Ka$$

$$pOH = - \log [OH^-]$$

are two examples.

177

2. Weak Acids and Bases.

Basically all weak acid–base problems are the same and we will illustrate the principles with a key example. What is the pH of a 0.010 M solution of the weak acid HA? The weak acid HA could be acetic acid, formic acid, or hydrocyanic acid for example. Let's say our weak acid HA has $K_a = 4.0 \times 10^{-6}$. The first thing to do in any of these calculations is write the equation and the equilbrium constant:

$$HA + H_2O \rightleftharpoons H_3O^+ + A^-$$

$$K_a = \frac{[H_3O^+][A^-]}{[HA]} \; .$$

Next make a table of initial concentrations and equilibrium concentrations

$$HA + H_2O \rightleftharpoons H_3O^+ + A^-$$

	HA	H_3O^+	A^-
initial	0.010	–	–
equilibrium	0.010 – x	x	x

In the above table x represents the amount of HA that ionizes in solution in moles liter^{-1}. Now substitute the equilbrium values in the K_a expression

$$K_a = \frac{[H_3O^+][A^-]}{[HA]} = \frac{(x)(x)}{(0.010-x)} = 4.0 \times 10^{-6} \; .$$

$$\frac{x^2}{0.010-x} = 4.0 \times 10^{-6}$$

$$x^2 + 4.0 \times 10^{-6} \, x - 4.0 \times 10^{-8} = 0$$

Using equation 12-12 in the text

$$x = \frac{-4.0 \times 10^{-6} \pm \sqrt{(4.0 \times 10^{-6})^2 - (4)(1)(-4.0 \times 10^{-8})}}{2}$$

$$= \frac{-4.0 \times 10^{-6} \pm \sqrt{16 \times 10^{-12} + 16 \times 10^{-8}}}{2}$$

$$= \frac{-4.0 \times 10^{-6} \pm \sqrt{16 \times 10^{-8}}}{2} \quad \text{(Note: } 16 \times 10^{-12} \text{ is much}$$
$$\text{smaller than } 16 \times 10^{-8})$$

$$= \frac{-4.0 \times 10^{-6} \pm 4.0 \times 10^{-4}}{2}$$

$$= \frac{-4.0 \times 10^{-6} + 4.0 \times 10^{-4}}{2} \quad \text{(The negative value has}$$
$$\text{no meaning).}$$

$$= 2.0 \times 10^{-4}$$

$$\therefore \quad [H_3O^+] = [A^-] = 2.0 \times 10^{-4}.$$

One can readily see that 2.0×10^{-4} is much smaller than the initial concentration of 0.010. Therefore one can avoid the tediousness of solving a quadratic equation by neglecting x with respect to 0.010. This gives

$$\frac{x^2}{0.010-x} = \frac{x^2}{0.010} = 4.0 \times 10^{-6}$$

$$\therefore x^2 = 4.0 \times 10^{-8}$$

$$x = 2.0 \times 10^{-4} \text{ as before.}$$

The question is—when can you neglect the amount of dissociation in order to allow the above simplification? A good rule of thumb to follow is the following: the approximation can be used if the initial concentration, C, is 10^{-2} or greater and K is 10^{-5} or smaller. Thus the magnitude of K is very important because it immediately gives you an idea of how much of the species dissociates. Thus a weak acid with a K_a of 1×10^{-2} would be considered to be a medium strength electrolyte and the quadratic equation must be used.

Problems involving weak bases are solved in exactly the same manner.

3. Buffers.

A buffer is a solution which resists changes in pH upon addition of small amounts of strong acid or base. A buffer contains both an acid and a base that can react with strong acids or bases to produce compounds which do not alter the pH from the value in the buffer. The most common buffer is composed of an acid and its conjugate base or a base and its conjugate acid. For example, for the weak acid HA, a buffer could be made up of the weak acid HA plus its conjugate base A^-:

$$\boxed{\begin{array}{l} HA, \\ A^- \end{array}}$$

This buffer can be prepared in two ways:

(1) mix approximately equal concentrations of HA and Na^+A^-.

(2) approximately half neutralize HA with strong base. i.e.

$$HA + OH^- \longrightarrow H_2O + A^-.$$

If we add H^+ (strong acid) to our buffer, $\boxed{HA, A^-}$, the acid will react with the basic component of the buffer:

$$\underset{\substack{\text{base in} \\ \text{buffer}}}{A^-} + \underset{\text{added}}{H^+} \longrightarrow HA$$

Thus we have used up the added strong acid by the above reaction producing HA (a weak acid) and the pH therefore does not change greatly.

180

If we add OH^- (strong base) to our buffer, $\boxed{HA, A^-}$, the base will

react with the acidic component of the buffer:

$$HA \qquad + \qquad OH^- \qquad \longrightarrow A^- + H_2O.$$

$$\text{acid in} \qquad \text{added}$$
$$\text{buffer}$$

In this case we have produced the conjugate base of the weak acid (a

weak base) and again the pH does not change appreciably. The reactions

written above essentially go to completion i.e. the equilibrium constants

are very large ($\sim 10^9$). Therefore the same principles as discussed in

chapter 3 are applicable. Solving buffer problems simply involves

finding the concentrations of HA and A^- and solving for $[H^+]$ using

the K_a expression. The same reasoning applies to basic buffers. We

will illustrate this in section B.

4. Titration of Acids and Bases

(a) Strong acid - Strong base.

In this case the net ionic equation of interest is

$$H^+ \quad + \quad OH^- \quad \underset{\leftarrow}{\rightarrow} \quad H_2O$$

$$\text{strong} \qquad \text{strong}$$
$$\text{acid} \qquad \text{base}$$

Since we know that $H_2O \underset{\leftarrow}{\rightarrow} H^+ + OH^-$, $K_w = 1.0 \times 10^{-14}$, then for

the reverse reaction $H^+ + OH^- \underset{\leftarrow}{\rightarrow} H_2O$, $K = \dfrac{1}{K_w} = 1.0 \times 10^{14}$. (Prove

this!) The large size of K means that the reaction goes to com-

pletion. The equivalence point in any titration is defined as

that point in the titration where stoichiometric amounts of acid

and base have been added. In the above titration we have only H_2O

produced at the equivalence point and therefore the solution is

<u>neutral</u> and the pH = 7.

(b) <u>Weak acid - Strong base</u>

For a weak acid, HA, the net ionic equation is HA + $OH^- \rightleftharpoons H_2O + A^-$.

For most weak acids the K for this reaction is large ($\sim 10^9$) and

the reaction goes to completion. In this case, however, the

products are H_2O and A^- at the equivalence point. Since A^- is

the conjugate base of the weak acid HA it causes hydrolysis of

water according to the equation

$$A^- + H_2O \rightleftharpoons HA + OH^-, K_h$$

and the resulting solution is basic. This can be understood by

noting that A^- wishes to react with H^+ from any source to form

HA. (The reaction $H^+ + A^- \rightleftharpoons HA$ has a large equilibrium constant equal

to $^1/_{K_a}$). The only source of H^+ ions is H_2O and thus hydrolysis

occurs to a small extent and A^- behaves as a weak base. The two

equilibria we have discussed are

$$H^+ + A^- \rightleftharpoons HA \qquad\qquad ^1/_{K_a}$$

$$H_2O \rightleftharpoons H^+ + OH^- \qquad\qquad K_w$$

If we add the above two equilibria we obtain

$$A^- + H_2O \rightleftharpoons HA + OH^- , \quad K_h .$$

When adding two equations the equilibrium constants are multiplied.

Therefore

$$K_h = \frac{K_w}{K_a} \text{ or } K_h K_a = K_w$$

You can verify this by writing out all the K's in full. For acetic

acid, CH_3CO_2H, plus NaOH we have $CH_3CO_2H + OH^- \rightleftharpoons CH_3CO_2^- + H_2O$. At

182

the equivalence point we essentially have a solution of $NaCH_3CO_2$ in water and thus in solution

$$CH_3CO_2^- + H_2O \rightleftharpoons CH_3CO_2H + OH^-$$

which represents the hydrolysis of water. In this example

$$K_h = \frac{K_w}{K_a} = \frac{1.0 \times 10^{-14}}{1.8 \times 10^{-5}} = 5.6 \times 10^{-10}.$$

Thus the solution produced at the equivalence point is basic due to hydrolysis of water.

(c) Weak base – Strong acid

For titration of the weak base NH_3 with the strong acid HCl the net ionic equation is

$$NH_3 + H^+ \rightleftharpoons NH_4^+.$$

At the equivalence point we have NH_4^+, a weak acid. The ammonium ion causes hydrolysis of water according to the equation

$$NH_4^+ + H_2O \rightleftharpoons NH_3 + H_3O^+$$

and the solution is acidic. (See Test 12-5).

The three types of titrations discussed above can be graphically illustrated by plotting the pH of a solution versus the volume of base added or the volume of acid added. The three types are illustrated on p. 184. In all three cases the equivalence point occurs when 50 ml of base ((a) and (b)) or 50 ml of acid ((c)) have been added. The discussion above is summarized in the table on page 185.

Titration Curves

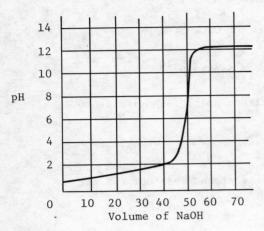

(a) Strong acid – Strong base.
(50 ml 0.10 M acid with 0.10 M base)

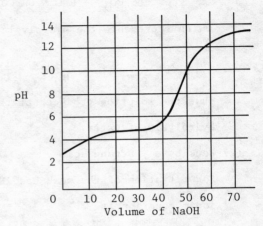

(b) Weak acid – Strong base.
(50 ml 0.10 M acid with 0.10 M base)

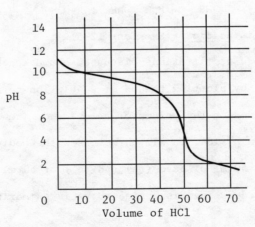

(c) Weak base – Strong acid.
(50 ml 0.10 M base with 0.10 M acid).

Type of Titration	Example	pH at the Equivalence Point
Strong acid – Strong base	$HCl + NaOH$	7.0
Weak acid – Strong base	$CH_3CO_2H + NaOH$	8.7
Weak base – Strong acid	$NH_3 + HCl$	5.3

An indicator is a reagent which changes color and signals the end-point in a titration. We hope to find out the equivalence point in a titration so we choose an indicator which changes color at the pH at the equivalence point (if possible). Since different titrations have different pH's at the equivalence point we also need different indicators. On page 186 the titration curves are shown, with various indicators and their color change pH range. It is clear from the figures that for (a) the "best" indicator is bromthymol blue since its color change occurs near the pH at the equivalence point. For (b) the best indicator is phenolphthalein and for (c) we would select methyl red. Remember, the indicator change is merely a signal and does not ordinarily affect the titration.

Finally, we noted previously that a buffer could be prepared by adding strong base to a weak acid (NaOH to CH_3CO_2H) until we had approximately half-neutralized the weak acid (CH_3CO_2H). This region of the curve, shown on the next page, is called the buffer region. It is easily seen that in this region the pH of the solution is not greatly altered by addition of small amounts of base. (curve is flat). The analogous region for the weak base titration is shown as well. It is also worthwhile pointing out that when exactly half of the acid has been neutralized we have equal amounts of HA and A^- in solution and therefore

(a) <u>INDICATORS</u>

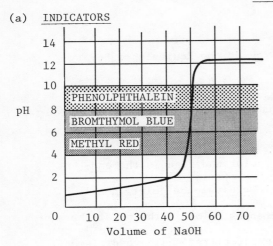

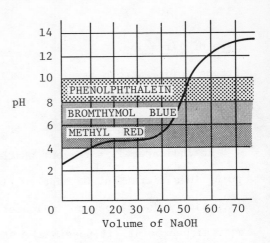

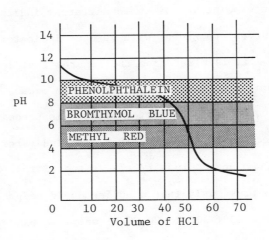

(b) <u>BUFFER REGION</u>

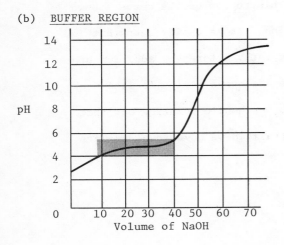

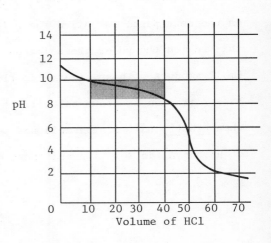

$$K_a = \frac{[A^-][H^+]}{[HA]} = [H^+].$$

This means that at half-neutralization (i.e. 25 ml NaOH) the pH = pK_a (or $[H^+] = K_a$).

B.   Types of Problems

Gaseous Equilibria

1.   Simply writing the equilibrium constant for any reaction.

2.   Given initial concentrations of products or reactants and one
     equilibrium concentration, calculate K.

3.   Given K and one or more initial concentrations, calculate the
     equilibrium concentrations.

4.   Apply Le Chatelier's Principle.

Aqueous Equilibria

1.   Calculations involving strong electrolytes and K_w.

2.   Weak acid and weak base ionization.

3.   Buffers.

4.   Titrations and hydrolysis.

Examples

Gaseous Equilibria

1.   Write the equilibrium constant expression for the reactions

$$H_2(g) + \frac{1}{8} S_8(s) \rightleftarrows H_2S(g)$$

and

$$CaCO_3(s) \rightleftarrows CaO(s) + CO_2(g).$$

Remembering not to include solids in the expression we obtain

$$K = \frac{[H_2S]}{[H_2]} \text{ and } K = [CO_2].$$

2.   Question 4 at the end of the chapter. You are given the following
     equation:

$$2A + B \rightleftarrows 3C + D$$

The initial concentration of A is equal to 1.8M and B, 1.8M. When equilibrium is attained the [C] is found to be 0.6. Calculate K, and indicate the units.

First tabulate the known information:

$$2A + B \rightleftarrows 3C + D$$

	2A	B	3C	D
initial	1.8	1.8	0	0
equilibrium			0.6	

One of the best procedures to follow in solving problems of this type is to find the reactant with coefficient one in the equation and allow x moles liter^{-1} to react to form products. In our case the reactant with coefficient one is B and if x moles liter^{-1} react to reach equilibrium then at equilibrium [B] = [initial] − [x] = 1.8 − x. For every B that reacts 3C and 1D must be formed. Also 2A's are lost for every B. This can be summarized as below:

$$2A + B \rightleftarrows 3C + D$$

	2A	B	3C	D
initial	1.8	1.8	0	0
equilibrium	1.8−2x	1.8−x	3x	x

In the problem we are given the value of 3x i.e.

$$3x = 0.6 \text{ M}$$

$$\therefore x = 0.2 \text{ M}$$

Therefore, at equilibrium [A] = 1.8−2x = 1.8−2(0.2)

$$= 1.4$$

$$[B] = 1.8-x = 1.8 - 0.2$$

$$= 1.6$$

189

$$[C] = 3x = 0.6$$

$$[D] = x = 0.2$$

Now write the equilibrium constant expression and substitute:

$$K = \frac{[C]^3 [D]}{[A]^2 [B]} = \frac{[0.6]^3 [0.2]}{[1.4]^2 [1.6]} = 0.01 \text{ M.} \quad \text{(1 significant figure).}$$

Always work your equilibrium problems with concentration units.

3. Into a 10.0 liter flask was introduced 2.000 mole of $COCl_2$. At $1000°C$ the equilibrium constant for the reaction

$$COCl_2(g) \rightleftarrows CO(g) + Cl_2(g)$$

is 0.329. Calculate the equilibrium concentration of all species at $1000°C$.

Let **x** M of $COCl_2$ react to form products at equilibrium. Therefore we have

$$COCl_2(g) \rightleftarrows CO(g) + Cl_2(g)$$

initial   2.000   0   0

equilibrium   2.000-x   x   x

$$\text{and } K = \frac{[CO][Cl_2]}{[COCl_2]} = \frac{x^2}{2.000-x} = 0.329$$

$$x^2 = 0.658 - 0.329x$$

$$x^2 + 0.329x - 0.658 = 0.$$

$$x = \frac{-0.329 \pm \sqrt{0.108 - 4(-0.658)}}{2}$$

$$= \frac{-0.329 \pm \sqrt{2.74}}{2}$$

$$= \frac{-0.329 \pm 1.66}{2}$$

= 0.665 (The negative solution has no meaning).

$\therefore$ [CO] = [Cl$_2$] = 0.665 and [COCl$_2$] = 2.000 − 0.665

= 1.34.

4. The methanation reaction in the gasification of coal is

$$CO(g) + 3H_2(g) \rightleftarrows CH_4(g) + H_2O(g).$$

Explain the effect (increase, decrease, no effect) on the number of moles of CH_4(g) produced when the following stresses are applied to the system. ΔH°_{rxn} = −49.3 kcal mole^{-1}.

(a) increase the concentration of H_2(g).

(b) increase the concentration of H_2O(g).

(c) remove some CH_4(g) from the system.

(d) increase the pressure (by decreasing the volume).

(e) addition of a catalyst.

(f) decrease the temperature.

(a) If one adds H_2(g) the equilibrium shifts so as to use H_2(g) and produce CH_4(g). Therefore the amount of CH_4(g) is increased.

(b) If H_2O(g) is increased the system shifts so as to use H_2O(g) and produce CO(g) and H_2(g). The amount of CH_4(g) is thus reduced.

(c) If CH_4(g) is removed, the system reacts so as to produce more CH_4(g), increasing the amounts of CO and H_2 consumed.

(d) An increase in pressure is relieved by shifting to the side of the equation that produces the smallest number of moles of gaseous species. Therefore the system is shifted to the right and CH_4(g) is increased.

(e) A catalyst does not alter the equilibrium and there is no effect on the CH_4(g) concentration.

191

(f) The reaction is <u>exothermic</u> (negative ΔH°_{rxn}) and thus an increase in temperature shifts the equilibrium to the left. A decrease in temperature shifts the equilibrium to the right. Therefore $CH_4(g)$ is increased by a decrease in temperature.

<u>Aqueous Equilibria</u>

1. (a) Calculate the pH of a solution in which $[OH^-] = 3.7 \times 10^{-3}$.

   (b) Calculate the $[H^+]$ of a solution whose pH = 4.74.

   (c) Calculate the pH of a solution prepared by mixing 50. ml of 0.020 M NaOH and 250 ml of 0.010 M HCl.

   Assume the volumes are additive. This is problem 12. at the end of the chapter.

   (a)

   $$K_w = [H^+][OH^-] = 1.0 \times 10^{-14}$$

   $$[H^+] = \frac{1.0 \times 10^{-14}}{[OH^-]} = \frac{1.0 \times 10^{-14}}{3.7 \times 10^{-3}} = 2.7 \times 10^{-12}$$

   $$pH = -\log [H^+]$$
   $$= -\log (2.7 \times 10^{-12})$$
   $$= -\{\log (2.7) + \log (10^{-12})\}$$
   $$= -\{0.43 + (-12)\}$$
   $$= 11.57$$

   (b)

   $$pH = -\log [H^+] = 4.74$$
   $$\log [H^+] = -4.74$$

   Rewrite -4.74 as a decimal fraction between 0 and 1 minus a whole number (always do this).

$$\log [H^+] = -4.74$$

$$= -5.00 + 0.26$$

$$\text{i.e. } -5.00 + 0.26 = -4.74$$

Now, $\log [H^+] = -5.00 + 0.26$

To obtain $[H^+]$ we need to take the antilog of both sides of the equation.

$$\text{antilog } (\log [H^+]) = \text{antilog } (-5.00 + 0.26)$$

$$[H^+] = \text{antilog } (-5.00) \times \text{antilog } (0.26)$$

$$= 10^{-5} \times 1.8 \qquad \swarrow\text{(when taking the antilog}$$
$$\text{of + you get x).}$$

$$\therefore [H^+] = 1.8 \times 10^{-5}$$

Again: What is the $[H^+]$ of a solution whose pH = 9.88?

$$pH = - \log [H^+]$$

$$\log [H^+] = -9.88$$

$$= -10.00 + 0.12$$

$$[H^+] = 10^{-10} \times 1.3$$

$$= 1.3 \times 10^{-10}.$$

Note the properties of logs in the Appendix.

(c) Write the equation for the reaction:

$$NaOH + HCl \rightarrow H_2O + NaCl$$

or the net ionic equation:

$$H^+ + OH^- \rightarrow H_2O$$

Calculate the number of moles of H^+ and OH^- and see what's left after the above reaction.

193

$$n_{H^+} = 0.010 \ \frac{mole}{\ell} \ x \ 0.250 \ \ell = 0.0025 \ mole$$

$$n_{OH^-} = 0.020 \ \frac{mole}{\ell} \ x \ 0.050 \ \ell = 0.0010 \ mole$$

One can see that all the OH^- reacts leaving

$$0.0025 - 0.0010 = 0.0015 \ mole \ H^+.$$

and $[H^+] = \dfrac{0.0015 \ mole}{0.300 \ \ell} = 5.0 \ x \ 10^{-3}.$

$pH = -log \ [H^+] = -log \ (5.0 \ x \ 10^{-3})$

$pH = 3.00 \ -log \ 5.0$

$\qquad = 3.00 - 0.70$

$pH = 2.30.$

2. (a) Question 21 at the end of the chapter.

Calculate the $[OH^-]$ for a 0.075 M solution of CH_3NH_2, methylamine
($K_b = 4.4 \ x \ 10^{-4}$).

Always write the equation. In this case we want K_b:

$$CH_3NH_2 + H_2O \rightleftharpoons CH_3NH_3^+ + OH^-$$

$$K_b = \frac{[CH_3NH_3^+][OH^-]}{[CH_3NH_2]}$$

Tabulate the data:

$$CH_3NH_2 + H_2O \rightleftharpoons CH_3NH_3^+ + OH^-$$

	CH_3NH_2	$CH_3NH_3^+$	OH^-
initial	0.075	—	—
equilibrium	0.075-x	x	x

$$K_b = 4.4 \ x \ 10^{-4} = \frac{x^2}{0.075-x}$$

In this example K_b is larger than 10^{-5} and the initial concentration,

0.075, is larger than 10^{-2} and thus we probably should solve the quadratic equation:

$$x^2 + 4.4 \times 10^{-4}x - 3.3 \times 10^{-5} = 0$$

$$x = 5.6 \times 10^{-3}$$

$$\therefore [OH^-] = 5.6 \times 10^{-3}.$$

The percent ionization is $\dfrac{5.6 \times 10^{-3}}{0.075} \times 100 = 7.5\%.$

This is greater than 5% and thus we would not have been justified in assuming x negligible with respect to 0.075.

(b) Question 23 at the end of the chapter.

The pH of vinegar is ≈ 3.0. Since vinegar is mainly a solution of acetic acid in water, calculate the concentration of undissociated acetic acid ($K_a = 1.8 \times 10^{-5}$).

Let acetic acid be represented by HA. Therefore

$$HA \quad + \quad H_2O \; \rightleftarrows \; H_3O^+ \; + \; A^-$$

initial $\quad C_{HA} \qquad\qquad\qquad - \qquad\quad -$

equilibrium $\quad C_{HA} - 1\times10^{-3} \qquad 1\times10^{-3} \quad 1\times10^{-3}$

$$K_a = \frac{[H_3O^+]\,[A^-]}{[HA]}$$

$$1.8\times10^{-5} = \frac{(1\times10^{-3})^2}{C_{HA} - 1\times10^{-3}}$$

$$1.8\times10^{-5}\,C_{HA} - 1.8 \times 10^{-8} = 1 \times 10^{-6}$$

$$C_{HA} = \frac{1\times10^{-6}}{1.8\times10^{-5}} = 0.06 \text{ M.}$$

(c) Question 29 at the end of the chapter.

Calculate the $[H^+]$ concentration of a solution made by mixing 200 ml of 0.50 M NaOH and 300 ml of 0.70 M HNO_2. ($K_a(HNO_2)$ = 4.5x10^{-4} from Table 12-1).

First write the equation for the reaction of NaOH with HNO_2: the net ionic equation is $OH^- + HNO_2 \longrightarrow H_2O + NO_2^-$.

Determine the number of moles of OH^- and HNO_2 and what is left over.

$$n_{OH^-} = 0.50 \frac{mole}{\ell} \times 0.200 \, \ell = 0.10 \text{ mole}$$

$$n_{HNO_2} = 0.70 \frac{mole}{\ell} \times 0.300 \, \ell = 0.21 \text{ mole}$$

	OH^-	+	HNO_2	$\longrightarrow$	H_2O	+	NO_2^-	
initial	0.10		0.21		–		–	} these are
								} moles.
final	–		0.11		–		0.10	}

Assume volumes are additive: $[HNO_2]$ = 0.22 $[NO_2^-]$ = 0.20

Since we have present in the solution a weak acid and its conjugat base we have a common ion or buffer problem. Now write the K_a expressic and summarize the above data:

	HNO_2	+	H_2O	$\rightleftharpoons$	H_3O	+	NO_2^-
initial (M)	0.22		–		–		0.20
equilibrium (M)	0.22-x				x		0.20 + x

$$4.5 \times 10^{-4} = \frac{x(0.20 + x)}{(0.22-x)}$$

K_a is larger than 10^{-5} but C_{HNO_2} and $C_{NO_2^-}$ are very much larger than 10^{-2} and we will probably be able to neglect x with respect to 0.20 and 0.22.

$$x = 4.5 \times 10^{-4} \times \frac{0.22}{0.20} = 5.0 \times 10^{-4}.$$

$$\therefore [H_3O^+] = 5.0 \times 10^{-4}. \text{ We can see that x is}$$

certainly negligible with respect to 0.20 or 0.22.

3. Question 37 at the end of the chapter.

(a) What is the pH of a solution made by adding 50.0 ml of 0.50 M sodium
acetate to 25. ml of 0.20 M HCl? (K_a (acetic acid) = 1.8×10^{-5})

(b) If 25. ml of 0.20 M HCl is added to the solution in (a), what is
the final pH?

(a) Write the equation for the reaction of sodium acetate and HCl: the
net ionic equation is (sodium acetate is $Na^+ CH_3 CO_2^-$).

$$CH_3 CO_2^- + H^+ \longrightarrow CH_3 CO_2 H$$

Determine the number of moles of $CH_3 CO_2^-$ and H^+ and what is left over
after reaction:

$$n_{CH_3 CO_2^-} = 0.050 \ \ell \times 0.50 \ \frac{mole}{\ell} = 0.025 \ \ mole$$

$$n_{H^+} = 0.025 \ \ell \times 0.20 \ \frac{mole}{\ell} = 0.005 \ mole.$$

$$CH_3 CO_2^- + H^+ \longrightarrow CH_3 CO_2 H$$

initial	0.025	0.005	–
final	0.020		0.005

these are moles.

We have a weak acid and its conjugate base in solution i.e. a buffer.

Assume volumes are additive to obtain concentrations:

$$CH_3 CO_2 H + H_2 O \rightleftharpoons CH_3 CO_2^- + H_3 O^+$$

initial, M	0.067	0.27	–
equilibrium, M	0.067–x	0.27 + x	x

$$1.8 \times 10^{-5} = \frac{(0.27 + x)(x)}{0.067 - x}$$

197

Since K is ~10^{-5} and $C_{CH_3CO_2H}$ and $C_{CH_3CO_2^-}$ are larger than 10^{-2} we can neglect x with respect to 0.27 and 0.067.

$$x = 1.8 \times 10^{-5} \times \frac{0.067}{0.27} = 4.5 \times 10^{-5}$$

$$[H_3O^+] = 4.5 \times 10^{-5}$$

$$pH = -\log [H^+]$$

$$= 5.00 - \log 4.5$$

$$= 5.35.$$

(b) Write the reaction which occurs when H^+ is added to a $CH_3CO_2H, CH_3CO_2^-$ buffer:

$$H^+ + CH_3CO_2^- \longrightarrow CH_3CO_2H$$

Determine the number of moles of all species:

$$n_{H^+} = 0.025 \; \ell \times 0.20 \; \frac{mole}{\ell} = 0.0050 \; mole.$$

$n_{CH_3CO_2^-} = 0.020$ moles (This is the final number of moles present from (a)).

Therefore

	H^+	$+ \; CH_3CO_2^-$	$\longrightarrow$	CH_3CO_2H	
initial	0.005	0.020		0.005	}
					} These are number of
final	–	0.015		0.010	} moles.

If volumes are additive $[CH_3CO_2^-] = \dfrac{0.015}{0.100} = 0.15$ and

$[CH_3CO_2H] = \dfrac{0.010}{0.100} = 0.10$ (Note we started with a volume of 75 ml and added 25 ml of acid).

Now solve:

	CH_3CO_2H	$+ \; H_2O$	$\rightleftharpoons$	$CH_3CO_2^-$	$+ \; H_3O^+$
initial	0.10	–		0.15	–
equilibrium	0.10-x	–		0.15+x	x

198

$$K_a = 1.8 \times 10^{-5} = \frac{(0.15 + x)(x)}{(0.10-x)}$$

Assume x is negligible with respect to 0.15 and 0.10.

$$x = 1.8 \times 10^{-5} \times \frac{0.10}{0.15} = 1.2 \times 10^{-5}$$

$$[H_3O^+] = 1.2 \times 10^{-5}.$$

$$pH = -\log [H^+]$$

$$= 5.00 - \log 1.2$$

$$= 4.92.$$

4.(i) Question 41 at the end of the chapter.

Write the equation for and calculate the value of the acid dissociation constant of pyridinium ion, $C_5H_5NH^+$.

Pyridine is a weak base and its K_b reaction is

$$C_5H_5N + H_2O \rightleftarrows C_5H_5NH^+ + OH^-$$

$$K_b = 1.4 \times 10^{-9}.$$

The pyridinium ion is a weak acid and dissociates slightly in water when added as a salt, say for example, as pyridinium chloride. The equation is

$$C_5H_5NH^+ + H_2O \rightleftarrows C_5H_5N + H_3O^+ , \quad K_h$$

$$K_h = \frac{K_w}{K_b} = \frac{1.0 \times 10^{-14}}{1.4 \times 10^{-9}} = 7.1 \times 10^{-6}.$$

One can see from the magnitude of K_h that the pyridinium ion is a much stronger acid than pyridine is a base.

(ii) Question 43 at the end of the chapter. Calculate the $[H^+]$ at the equivalence point of the titration of 0.25M NH_3(aq) with 0.25M hydrochloric acid.

199

Write the equation for the reaction.

$$NH_3 + H^+ \longrightarrow NH_4^+$$

At the equivalence point, we have added a number of H^+ ions equal to the number of NH_3 molecules originally present. Since they react in a 1:1 ratio, $[NH_3] = [H^+]$, which is just the situation if the solution were made from the same number of NH_4^+ ions. Since the original NH_3 and HCl concentrations are equal, the volume at the equivalence point is twice the volume of the original NH_3 solution. Thus, the equivalence point solution can be regarded as a solution of NH_4^+ at half the original NH_3 concentration. Therefore the concentration of NH_4^+ is 0.25 M ÷ 2 = 0.13 M. If you don't "see" that the NH_4^+ concentration is one half of the starting concentration, assume a volume of NH_3 to start and go through the calculation. i.e. assume 1.0 ℓ of 0.25 M NH_3 is titrated with 0.25 M HCl. Since NH_4^+ acts as a weak acid we have

$$NH_4^+ + H_2O \rightleftarrows NH_3 + H_3O^+$$

	NH_4^+	H_2O	NH_3	H_3O^+
initial	0.13	–	–	–
equilibrium	0.13-x		x	x

$$K_h = \frac{x^2}{0.13-x}$$

In Table 12-1 we are given K_b for $NH_3 = 1.8 \times 10^{-5}$.

$$\therefore \quad K_h = \frac{K_w}{K_b} = \frac{1.0 \times 10^{-14}}{1.8 \times 10^{-5}} = 5.6 \times 10^{-10}$$

$$\therefore \quad 5.6 \times 10^{-10} = \frac{x^2}{0.13-x}$$

Since K is much smaller than 10^{-5} and the initial concentration, $C_{NH_4^+}$, is larger than 10^{-2} we can neglect x with respect to 0.13.

$$x^2 = 0.13 \times 5.6 \times 10^{-10}$$

$$= 0.73 \times 10^{-10}$$

$$x = 0.85 \times 10^{-5}$$

$$[H_3O^+] = 8.5 \times 10^{-6}$$

$$\text{and pH} = 5.07$$

C. Test Yourself

1. You are given the following equilibrium

$$A + 3B \rightleftharpoons 2C$$

For initial concentrations of [A] = 1.00, [B] = 3.00 M, and [C] = 0, equilibrium is attained and the concentration of C is found to be 0.980 M. Calculate K.

2. State the direction in which the following equilibria would be shifted upon application of the stress listed beside the equation, i.e. right, left, no change.

(a) $CaCO_3(s) \rightleftharpoons CaO(s) + CO_2(g)$

        remove CO_2.

(b) $CO(g) + H_2O(g) \rightleftharpoons CO_2(g) + H_2(g)$

        add H_2O.

(c) $3Fe(s) + 4H_2O(g) \rightleftharpoons Fe_3O_4(s) + 4H_2(g)$

    decrease total pressure by increasing volume.

3. Calculate the pH of a solution prepared by mixing 200. ml of 0.30 M $HClO_4$ and 300 ml of 0.050 M $Ca(OH)_2$.

4. Calculate the pH of a 0.10 M solution of HF. (K_a for HF = 7.2×10^{-4}).

5. Calculate the pH of a 0.10 M solution of NaF. (K_a for HF = 7.2×10^{-4}).

6. Calculate the weight in grams of ammonium chloride (NH_4Cl) which would have to be added to 500.0 ml of 0.500 M NH_3 to give a pH of 9.00.

7. Ethylamine, $C_2H_5NH_2$, is a weak base having a $K_b = 5.6 \times 10^{-4}$. Calculate the pH at the equivalence point in a titration of 0.240 M aqueous ethylamine with 0.240 M aqueous HCl.

202

8. The K_a of β-chlorobutyric acid, $HC_4H_6ClO_2$, is 8.9×10^{-5}. Calculate the weight of β-chlorobutyric acid which must be added to 5.00 liter of water to yield a solution of pH 3.00. Assume no change in volume.

9. The $[H^+]$ in a buffer solution prepared by dissolving 0.240 mole of butyric acid, $HC_4H_7O_2$, and 0.120 mole of sodium butyrate, $NaC_4H_7O_2$, in enough water to yield 1.00 liter of solution is 3.00×10^{-5} M. Calculate the K_a of butyric acid.

10. Calculate the pH of a solution prepared by adding 4.00 g NaOH to 500.0 ml of 1.0 M HOCN. Assume no volume change. (K_a for HOCN = 2.2×10^{-4}).

11. (a) Write a balanced net ionic equation for the reaction which occurs when nitric acid is added to an ammonia-ammonium chloride buffer.

   (b) Write a balanced net ionic equation for the reaction which occurs when NaOH is added to a pyridine-pyridinium chloride buffer.

13

Heterogeneous Equilibria

A. Points of Importance

The Solubility Product

The solubility product is the equilibrium constant for the solubility in water of a relatively insoluble ionic solid. For example, for $Al(OH)_3$ we have

$$Al(OH)_3(s) \rightleftharpoons Al^{3+} + 3OH^- \quad \text{and} \quad K_{sp} = [Al^{3+}][OH^-]^3.$$

For $PbSO_4$ the equation is

$$PbSO_4(s) \rightleftharpoons Pb^{2+} + SO_4^{2-} \quad \text{and} \quad K_{sp} = [Pb^{2+}][SO_4^{2-}].$$

The solubility of $PbSO_4(s)$ in water is 1.3×10^{-4} M therefore the

$$
\begin{aligned}
K_{sp} &= [Pb^{2+}][SO_4^{2-}] \\
&= (1.3 \times 10^{-4})(1.3 \times 10^{-4}) \\
&= 1.7 \times 10^{-8}.
\end{aligned}
$$

Application of Le Chatelier's principle predicts that the addition of SO_4^{2-}

to the solution will result in a lower concentration of Pb^{2+} and formation of $PbSO_4$.

The value of K_{sp} can be used to predict whether or not a precipitate will form when various concentrations of ions are mixed.

(1) If K_{sp} > initial ion product $\longrightarrow$ no precipitation.

(2) If K_{sp} < initial ion product $\longrightarrow$ precipitation occurs until the product of the concentrations of ions in solution = K_{sp}.

B. Underline: Types of Problems

1. Given K_{sp} calculate the solubility of the salt.

2. Given the solubility of the salt, calculate K_{sp}.

3. Common ion problems involving K_{sp}.

Examples

1.(i) Calculate the solubility of iron(II) hydroxide given K_{sp} = 1.6 x 10^{-15}. Let the solubility be x.

The equation is

$$Fe(OH)_2(s) \rightleftharpoons Fe^{2+} + 2OH^-$$

initial   0   0

equilibrium   x   2x

$$K_{sp} = [Fe^{2+}][OH^-]^2$$
$$1.6 \times 10^{-15} = (x)(2x)^2$$
$$4x^3 = 1.6 \times 10^{-15}$$
$$x^3 = 0.4 \times 10^{-15}$$
$$x = 7.4 \times 10^{-6}.$$

(ii) Question 14 at the end of the chapter.

A solution contains 1 x 10^{-3} M magnesium ion. If enough NaOH is added to make the $[OH^-]$ equal to 1 x 10^{-5} M, will precipitation occur? (K_{sp} = 1.1 x 10^{-11}).

$$Mg(OH)_2(s) \rightleftharpoons Mg^{2+} + 2OH^-$$

$$[Mg^{2+}][OH^-]^2 = (1 \times 10^{-3})(1 \times 10^{-5})^2 = 1 \times 10^{-13}$$

$$K_{sp} = 1 \times 10^{-11}$$

Since $K_{sp} > [Mg^{2+}][OH^-]^2$ no precipitation occurs.

2.   Question 15 at the end of the chapter.

At 25°C, 0.160 g of $SrSO_4$ dissolve per liter of water.

Calculate K_{sp}.

$$\text{Solubility} = \frac{0.160 \text{ g}}{\ell} \times \frac{1 \text{ mole}}{183.68 \text{ g}} = 8.71 \times 10^{-4} \frac{\text{mole}}{\ell}$$

$$SrSO_4(s) \rightleftarrows Sr^{2+} + SO_4^{2-}$$

$$K_{sp} = [Sr^{2+}] [SO_4^{2-}]$$

$$= (8.71 \times 10^{-4}) (8.71 \times 10^{-4})$$

$$= 7.59 \times 10^{-7}.$$

3.   Question 17 at the end of the chapter.

How many moles of MgF_2 will dissolve in a liter of 0.010 M NaF?
(For simplicity neglect 2x compared to 0.010 in this problem).
$K_{sp} = 8 \times 10^{-8}$.

$$MgF_2(s) \rightleftarrows Mg^{2+} + 2F^-$$

initial	0	0.010
equilibrium	x	0.010 + 2x

$$K_{sp} = [Mg^{2+}][F^-]^2$$

$$8 \times 10^{-8} = (x)(0.010 + 2x)^2$$

$$x = \frac{8 \times 10^{-8}}{0.010} = 8 \times 10^{-6}.$$

Therefore 8×10^{-6} moles of MgF_2 will dissolve in a liter of
0.010 M NaF.

C. Test Yourself

1. Calculate the solubility of $Pb_3(PO_4)_2$ in water. $(K_{sp} = 1 \times 10^{-42})$.

2. Calculate the solubility of BaF_2 in water. $(K_{sp} = 2 \times 10^{-6})$.

3. If 500.0 ml of 0.10 M $NaClO_4$ are added to 500.0 ml of 0.20 M KCl will $KClO_4$ precipitate from solution? K_{sp} $(KClO_4) = 2.1 \times 10^{-2}$. If not, what concentration of $NaClO_4$ (500 ml) must be added to just cause precipitation?

4. The solubility of CaF_2(s) in H_2O is 2×10^{-5} M. Calculate K_{sp}.

5. How many moles of $BaSO_4$ will dissolve in a liter of 0.010 M Na_2SO_4? $K_{sp} = 1.5 \times 10^{-9}$.

6. If solid $AgClO_4$ is added to a solution containing 0.050 M $C_2O_4^{2-}$ and 0.050 M Cl^-, which will precipitate first, $Ag_2C_2O_4$ or AgCl? K_{sp} $(Ag_2C_2O_4) = 5.3 \times 10^{-12}$ and K_{sp} (AgCl) $= 1.6 \times 10^{-10}$.

7. Will a precipitate form if 25.0 ml of 0.0010 M $AgClO_4$ and 25.0 ml of 0.0010 M NaSCN are mixed? K_{sp} (AgNCS) $= 1.0 \times 10^{-12}$.

8. Which species has the highest solubility, $PbCO_3$ or $Ag_2C_2O_4$? K_{sp} $(PbCO_3) = 1 \times 10^{-13}$ and K_{sp} $(Ag_2C_2O_4) = 5.3 \times 10^{-12}$.

14

The Second Law of Thermodynamics

A.   Points of Importance

Free Energy (See sections 14-1 and 14-2 in the text).

Free energy changes are commonly expressed for standard states.
The standard free energy of formation, ΔG_f°, is defined as the free
energy change of the reaction in which the compound is formed in its
standard state from its elements in their standard states. The ΔG° for
a reaction where reactants in their standard states are converted to
products in their standard states can be calculated from:

$$\Delta G^\circ = \Sigma n_p \Delta G_f^\circ \text{ (products)} - \Sigma n_r \Delta G_f^\circ \text{ (reactants)}.$$

The standard free energy of a reaction is also related to the equilibrium constant for the reaction:

$$\boxed{\Delta G^\circ = -RT\ln K.}$$

Thus the sign of $\Delta G°$ determines the feasibility of a reaction i.e. whether products or reactants are favored.

For standard conditions $\Delta G°$ is related to $\Delta H°$ by the equation

$$\Delta G° = \Delta H° = T\Delta S°$$

where $\Delta S°$ is the standard entropy of the system (related to randomness and probability as described in the text). The sign of $\Delta G°$ in this equation depends on the sign and magnitude of $\Delta H°$ and $\Delta S°$ as shown in the table on the next page.

Practical use of a Table of this type is found in the purification of Nickel by the Mond process:

(1) $Ni(s)$ + $4CO(g)$ $\xrightarrow{\;80°C\;}$ $Ni(CO)_4(g)$
       impure

(2) $Ni(CO)_4(g)$ $\xrightarrow{\;200°C\;}$ $Ni(s) + 4CO(g)$
                                                      pure

For reaction (1) the entropy of formation of $Ni(CO)_4(g)$ is negative as is the enthalpy of formation. The negative $\Delta H_f°$ term is larger than the $-T\Delta S°$ term (which is positive) and $\Delta G_f°$ is negative indicating the reaction will occur at room temperature. However as the temperature is increased the positive $T\Delta S°$ term becomes larger and $\Delta G_f°$ becomes positive (for reaction (1)). This means that reaction (2) (opposite of (1)) will occur at high temperature. One can thus produce pure $Ni(CO)_4$ at low temperatures leaving behind impurities and then deposit pure Ni at higher temperatures.

Sign of $\Delta H°$	Sign of $\Delta S°$	Sign of $\Delta G°$ ($\Delta G° = \Delta H° - T\Delta S°$)
positive	negative	positive at all temperatures
negative	positive	negative at all temperatures
positive	positive	depends on magnitude of $\Delta H°$ and $\Delta S°$ and on the temperature. At high temperatures it is likely that $- T\Delta S° > \Delta H°$ and thus $\Delta G°$ would be negative.
negative	negative	depends on magnitude of $\Delta H°$ and $\Delta S°$ and on the temperature. At high temperatures it is likely that $-T(-\Delta S°) > -\Delta H°$ and thus $\Delta G°$ would be positive.

B. Types of Problems

1. Calculation of $\Delta G°$ from $\Delta G_f°$ data and prediction of the feasibility of a reaction.

2. Calculations involving the equation $\Delta G° = -RT\ell n K$.

3. Calculations involving the equation $\Delta G° = \Delta H° - T\Delta S°$.

4. Use of absolute entropies to calculate $\Delta S°$. The equation

$$\Delta S° = \Sigma n_p S° \text{ (products)} - \Sigma n_r S° \text{ (reactants)}.$$

Examples

1. Question 14 at the end of the chapter.

   In which direction is the following reaction feasible?

   $\Delta G_f°(H_2S(g)) = -7.9$ kcal mole^{-1}.

   $$8H_2(g) + S_8(s) \rightleftharpoons 8H_2S(g).$$

   $$\Delta G° = \Sigma n_p \Delta G_f° \text{ (products)} - \Sigma n_r \Delta G_f° \text{ (reactants)}$$

   $$= 8 \times -7.9 - 0$$

   $$= -63 \text{ kcal mole}^{-1}$$

   Since $\Delta G°$ is negative the position of equilibrium lies to the right and the reaction from left to right is feasible.

2. Question 11 at the end of the chapter.

   Calculate K from the data in Table 14-1 for the reaction:

   $$PCl_3(g) + Cl_2(g) \rightleftharpoons PCl_5(g)$$

   $$\Delta G° = [-77.6] - [-68.4]$$

   $$= -9.2 \text{ kcal mole}^{-1}.$$

   But $\Delta G° = -RT\ell n K$

$$\ell nK = - \frac{\Delta G^\circ}{RT}$$

$$\ell nK = - \frac{(-9.2 \times 10^3 \text{cal mole}^{-1})}{(1.987 \text{ cal mole}^{-1}{}^\circ K^{-1})298^\circ K}$$

$$\ell nK = 15.5$$

$$2.303 \log K = 15.5$$

$$\log K = 6.73$$

$$K = 5.4 \times 10^6.$$

3. (i) Question 15 at the end of the chapter.

The equilibrium constant for the reaction

$$H_2(g) + I_2(g) \rightleftharpoons 2HI(g)$$

is 66.5 at 633°K and 50.7 at 713°K.

(a) Is the reaction exothermic?

(b) Calculate ΔH°.

(a) In this example the equilibrium constant has decreased as a result
     of adding heat (increase in temperature). This means that the
     reaction has been shifted to the left as a result of the "applied
     stress", an increase in temperature. Let's apply Le Chatelier's
     principle to this situation. First write the reaction as both an
     <u>endothermic</u> and an <u>exothermic</u> reaction.

(1) heat + $H_2(g) + I_2(g) \rightleftharpoons 2HI(g)$ (endothermic)

(2) $H_2(g) + I_2(g) \rightleftharpoons 2HI(g) +$ heat   (exothermic)

     For (1) Le Chatelier's principle predicts that addition of heat
     will shift the equilibrium to the right. This is contrary to the
     given information (shifts to the left) and therefore the reaction

213

cannot be endothermic. For (2) addition of heat shifts the equilibrium to the left and thus the reaction is <u>exothermic</u>.

(b)   To calculate $\Delta H°$ we use the equation (14-8):

$$\ell nK = \frac{-\Delta H°}{RT} + \frac{\Delta S°}{R}$$

$$2.303 \; logK = \frac{-\Delta H°}{RT} + \frac{\Delta S°}{R}$$

$$logK = \frac{-\Delta H°}{2.303R} \cdot \frac{1}{T} + \frac{\Delta S°}{2.303R}$$

We can plot logK versus $\frac{1}{T}$ and obtain $\Delta H°$ from the slope of the line. (Since we only have two points we should manage an excellent straight line). The data are summarized below:

T(°K)	K	$\frac{1}{T}$(°K)$^{-1}$	log K
633	66.5	1.58×10^{-3}	1.823
713	50.7	1.40×10^{-3}	1.705

Taking the two points above we can obtain the slope as:

$$slope = \frac{rise \; (y)}{run(x)} = \frac{1.823 - 1.705}{(1.58 - 1.40) \times 10^{-3}} \; °K$$

$$= 0.656 \times 10^3 \; °K$$

But the slope $= \frac{-\Delta H°}{2.303R} = 0.656 \times 10^3 \; °K$

$$\therefore \quad \Delta H° = -2.303 \times 0.656 \times 10^3 \; °K \times \frac{1.987 \; cal}{mole \; °K}$$

$$= -3.00 \times 10^3 \; cal \; mole^{-1}$$

$$\Delta H° = -3.00 \; kcal \; mole^{-1}$$

If $\Delta S°$ were desired we could calculate it from equation 14-8 by substitution of K at a given T along with $\Delta H°$. This procedure is easiest since a very large extrapolation to $\frac{1}{T} = 0$ is necessary to

214

obtain the intercept graphically.

3.(ii) Which of the two reactions below would you expect to have the greatest $\Delta S°$ value?

$$2CO(g) + 2NO(g) \rightleftharpoons 2CO_2(g) + N_2(g)$$

or

$$SF_6(g) \rightleftharpoons SF_4(g) + F_2(g).$$

The second reaction would be expected to have the greater $\Delta S°$ value. The $\Delta S°$ value for this reaction is expected to be positive because a collection of SF_6 molecules is more highly ordered than an equal amount of SF_4 and F_2 molecules. For the first reaction there is more order in the products since there are only 3 moles of gas whereas in the reactants we have 4 moles of gas and greater disorder. For this reaction we would expect $\Delta S°$ to be negative.

3.(iii) For the reaction below $\Delta H° = + 21.0$ kcal mole^{-1} and $\Delta S° = + 40.7$ cal mole^{-1}deg^{-1}:

$$PCl_5(g) \rightleftharpoons PCl_3(g) + Cl_2(g)$$

(a) Is this reaction feasible at 25°? If not, would you raise or lower the temperature in an attempt to make it feasible?

(b) Above what temperature would this process be expected to be feasible?

(a) Calculate $\Delta G°$ to answer this question.

$$\Delta G° = \Delta H° - T\Delta S° \qquad (T = 298°K \text{ for standard state})$$

Convert $\Delta S°$ to kcal mole^{-1}°K^{-1}.

$$\Delta G° = 21.0 - 298°K \times 40.7 \times 10^{-3} \text{ kcal mole}^{-1}°K^{-1}$$

$$= 21.0 - 12.1$$

215

$$= +8.9 \text{ kcal mole}^{-1}.$$

Since $\Delta G°$ is positive the reaction is not feasible.

Since $\Delta H°$ is positive and $\Delta S°$ is positive we must raise the temperature until the value of $-T\Delta S°$ exceeds $\Delta H°$. At this point $\Delta G°$ will become negative.

(b) At $\Delta G° = 0$ we can calculate T from $\Delta G° = \Delta H° - T\Delta S°$

Above this temperature $\Delta G°$ must be negative and the process feasible.

$$\Delta G° = 0 = \Delta H° - T\Delta S°$$

$$T = \frac{\Delta H°}{\Delta S°} = \frac{21.0 \text{ kcal mole}^{-1}}{40.7 \times 10^{-3} \text{kcal mole}^{-1}°\text{K}^{-1}}$$

$$= 516°\text{K}$$

4. Calculate $\Delta S°$ at 25°C for the following reactions:

(a) $C(s) + O_2(g) \rightleftarrows CO_2(g)$

(b) $H_2O_2(\ell) \rightleftarrows H_2O(\ell) + \frac{1}{2}O_2(g)$

(c) $N_2O_4(g) \rightleftarrows 2NO_2(g)$

$S°(C(s)) = 1.4$, $S°(O_2(g)) = 49.0$, $S°(CO_2(g)) = 51.1$,

$S°(H_2O_2(\ell)) = 26.2$, $S°(H_2O(\ell)) = 16.7$, $S°(N_2O_4(g)) = 72.7$,

$S°(NO_2(g)) = 57.5 \text{ cal mole}^{-1}\text{deg}^{-1}$.

(a) $\Delta S° = \Sigma n_p S°$ (products) $- \Sigma n_r S°$ (reactants)

$$= 51.1 - [1.4 + 49.0]$$

$$= + 0.7 \text{ cal mole}^{-1}°\text{K}^{-1}$$

(b) $\Delta S° = [16.7 + \frac{1}{2} \times 49.0] - [26.2]$

$$= + 15.0 \text{ cal mole}^{-1}°\text{K}^{-1}$$

(c) $\Delta S° = 2 \times 57.5 - 72.7$

$$= 115.0 - 72.7$$

$$= + 42.3 \text{ cal mole}^{-1}°\text{K}^{-1}.$$

216

C.   Test Yourself

1.(a) Using the data in Table 14-1 calculate $\Delta G°$ for the reaction below

   proposed for use in a catalytic converter on automobiles:

$$2CO(g) + 2NO(g) \rightleftarrows 2CO_2(g) + N_2(g)$$

   (b) Is this reaction feasible?

   (c) Would you expect $\Delta S°$ for this reaction to be positive or negative?

   Confirm your answer using the data in Table 14-3.

   (d) Calculate $\Delta H°$ for this reaction.

2.   For the reaction

$$C_2N_2(g) + O_2(g) \rightleftarrows 2CO(g) + N_2(g)$$

   the following data were calculated:

T(°C)	K	log K
-175	8.6×10^{289}	289.93
-75	1.4×10^{141}	141.14
25	1.9×10^{100}	100.28
125	$6.7 \times 10^{7.6}$	76.83
225	7.8×10^{62}	62.89

   (a) Determine $\Delta H°$ and $\Delta S°$.

   (b) Calculate $\Delta G°$ from K at 25°C and from the data in Table 14-1

   $(\Delta G_f° (C_2N_2(g)) = 71.1 \text{ kcal mole}^{-1})$.

3.   Given $\Delta H° = -10.0 \text{ kcal mole}^{-1}$ and $\Delta S° = -5.2 \text{ cal mole}^{-1}°K^{-1}$

   for the reaction

$$SO_2(g) + NO_2(g) \rightleftarrows SO_3(g) + NO(g).$$

   (a) Is the reaction feasible at room temperature?

217

(b) Will the extent of reaction increase or decrease at higher

temperatures?

4.(a) Would you expect the reaction below to have a positive or a

negative $\Delta S°$? Explain.

$$N_2H_4(\ell) + 2H_2O_2(\ell) \rightleftarrows N_2(g) + 4H_2O(g)$$

(b) If $\Delta H°$ for this reaction is -126.7 kcal mole^{-1} and using your

answer from part (a) would you expect the extent of the reaction

to increase or decrease at higher temperatures?

5.   For the reaction

$$SF_6(g) \rightleftarrows SF_4(g) + F_2(g)$$

$\Delta S° = 48.5$ cal mole^{-1}deg^{-1} and $\Delta H° = 104$ kcal mole^{-1}.

(a) Calculate $\Delta G°$ and the equilibrium constant at 25°C.

(b) Is this reaction feasible as written? Explain.

(c) At what temperature would the reaction become feasible?

6.   The standard free energies of formation of $C_2H_6(g)$, $C_2H_4(g)$, and

$C_2H_2(g)$ are -7.9, $+16.3$, and $+50.0$ kcal mole^{-1}.

(a) What does this imply about the preparation of these compounds from

their elements?

(b) What does this imply about their potential use as fuels (i.e. their

combustion)?

7.   Calculate $\Delta G°$ for the "roasting" of $PbCO_3(s)$

$$PbCO_3(s) \longrightarrow PbO(s) + CO_2(g).$$

Under what conditions will this reaction be feasible?

($\Delta H° = +20.8$ kcal mole^{-1}, $\Delta S° = +36$ cal mole^{-1}deg^{-1}).

15

Interconversion of Chemical and Electrical Energies

A.   Points of Importance

1.   New Terms

Memorize the following terms and their definitions.

Oxidation is the loss of electrons: $Cs(s) \rightarrow Cs^+(aq) + e^-$.

(This is shown by the electron on the right hand side).

Reduction: is the gain of electrons: $Co^{2+}(aq) + 2e \rightarrow Co(s)$.

(This is shown by the electrons on the left hand side).

Anode: the electrode where oxidation occurs. For voltaic cells the anode is negative. For electrolytic cells the anode is positive.

Cathode: the electrode where reduction takes place.

For voltaic cells the cathode is positive. For electrolytic cells the cathode is negative.

Hydrogen electrode: standard half-cell, $2H^+ + 2e^- \rightarrow H_2$ arbitrarily assigned a voltage of zero.

Voltaic cell: a cell which produces electricity from a chemical reaction that occurs spontaneously.

Electrolytic cell: a cell in which energy is used to cause a non-spontaneous chemical reaction to occur.

The Faraday: 1 mole of electrons = 96,500 coulombs. Also, $1\ F = 23,060$ calories volt^{-1}.

2. Design of Voltaic Cells

Electrochemical cells can be designed by use of the data in Table 15-3. For example we could construct a voltaic cell using the two half-reactions.

(1) $Cd^{2+}(aq) + 2e^- \longrightarrow Cd(s)$ $\qquad\qquad$ $E° = -0.40$ v

(2) $Cl_2(g) + 2e^- \longrightarrow 2Cl^-(aq)$ $\qquad\qquad$ $E° = +1.36$ v

The half-reaction with the more positive potential, (2), will occur as written and the other half-reaction will occur in reverse. The overall reaction and $E°$ is obtained by adding the reactions:

$$Cl_2(g) + 2e^- \longrightarrow 2Cl^-(aq) \qquad\qquad E° = +1.36 \text{ v}$$

$$\underline{Cd(s) \qquad\quad \longrightarrow Cd^{2+}(aq) + 2e^- \qquad\qquad E° = +0.40 \text{ v}}$$

$$Cd(s) + Cl_2(g) \longrightarrow Cd^{2+}(aq) + 2Cl^- \qquad\qquad E° = +1.76 \text{ v}$$

Now we can draw a voltaic cell in which this reaction occurs to do some form of useful work. Since the E° of the overall reaction is positive the reaction occurs. First mark on the diagram on the next page the cathode (where reduction occurs) and the anode (where oxidation occurs). Next put the proper reagents in the proper compartments. Mark the direction of flow of electrons –

220

electrons always flow toward the positive terminal (the anode in this case). Also show the direction of flow of the cations and the anions - the cations flow toward the cathode and the anions toward the anode. The salt bridge simply contains an inert electrolyte such as NaCl to maintain electroneutrality.

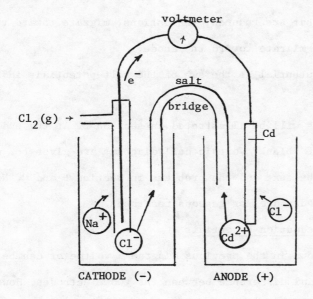

Remember the cathode reaction is always the half-reaction which has the largest reduction potential i.e. the furthest to the bottom in Table 15-3.

The voltmeter in the voltaic cell measures the potential difference between the cells. In this case this potential is 1.76 v.

Generally you will be asked to design voltaic cells <u>given the</u> overall reaction. The following rules should be followed:

221

(1) Write the two half-reactions as reduction potentials and indicate their potentials.

(2) The couple with the more positive potential occurs as written (reduction at cathode) and the other reaction occurs in reverse (oxidation at anode).

(3) Electrons travel toward the positive electrode – the cathode.

(4) The ions that are reduced, the cations, migrate toward the cathode.

(5) The anions migrate toward the anode.

(6) The cell potential is the sum of the half-potentials indicated in (2).

These rules will be illustrated in section B. At the end of this sec a number of "blank" voltaic cell diagrams are given for use in doing tests in the text and the problems in Section B and C. Note that som the electrodes are for gaseous reactants.

3.   Standard Reduction Potentials

As shown in the previous diagram a voltmeter can be used to mea the potential difference berween the two electrodes. However, we car measure the potential of a half reaction separately and it must be measured with respect to a reference electrode. The reference electr is the hydrogen electrode: $2H^+ + 2e \rightarrow H_2$, which has a defined potent of zero. The potentials or emf of other half reactions are then measured against the hydrogen electrode. For example we can use a voltaic cell to measure the potential of the half reaction $Ag^+(aq) + e^- \rightarrow Ag(s)$ by using the hydrogen electrode as a reference At 25°C, when the concentrations of $H^+(aq)$ and $Ag^+(aq)$ are 1M and

222

the pressure of $H_2(g)$ is 1 atmosphere the voltmeter reads 0.80 V.
The direction of the deflection of the voltmeter indicates that Ag(s)
has a smaller tendency to give off electrons than H_2. In other words
the reduction of Ag^+ is favoured. This means that the half reaction
$Ag^+(aq) + e^- \rightarrow Ag(s)$ occurs at the cathode. Thus the hydrogen half-
reaction is oxidation at the anode:

$$H_2(g) \rightarrow 2H^+(aq) + 2e^-.$$

The overall reaction is the sum of the two half-reactions:

$$2 \times \quad [Ag^+(aq) + e^- \rightarrow Ag(s)] \qquad\qquad E° = 0.80 \text{ v}$$

$$H_2(g) \xrightarrow{\quad\quad} 2H^+(aq) + 2e^- \qquad\qquad E° = 0.00 \text{ v}$$

$$2Ag^+(aq) + H_2(g) \longrightarrow 2Ag(s) + 2H^+(aq) \quad E° = 0.80 \text{ v}$$

Note in the above equations we multiplied the Ag^+/Ag half-reaction by
2 but not the E° for this reaction. Since the overall potential can be
measured and since the $H_2/2H^+$ couple has an E° = 0.00 the value + 0.80 v
represents the reduction potential for the Ag^+/Ag half-reaction. This
voltaic cell is shown in the diagram on the next page. In this cell
A^- simply represents any inert anion from the added electrolyte eg.
SO_4^{2-}, NO_3^- from Na_2SO_4 or $NaNO_3$. Try constructing a voltaic cell to
measure the E° for the half-reaction $Zn^{2+}(aq) + 2e^- \rightarrow Zn(s)$. In this
case the direction of deflection of the voltmeter indicates that Zn(s)
has a larger tendency to give off electrons than H_2. Does this make
the Zn/Zn^{2+} half-reaction at the anode or cathode? (anode!)

223

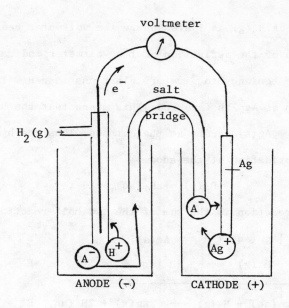

voltmeter

e⁻

salt bridge

$H_2(g) \rightarrow$

Ag

A^-

A^-

H^+

Ag^+

ANODE (−)   CATHODE (+)

Table 15-3 is constructed in this manner. The higher the posit

value the greater the tendency for reduction to occur. The higher t

negative value, the greater the tendency for oxidation to occur.

Remember the best reducing agent is in the upper right. Thus when th

reduced form of an element is combined with the oxidized form of an

element below it in the table, spontaneous reaction will occur. This

means that the cell potential for the reaction, E° (obtained by addi

the appropriate half-reactions), is positive.

4. Electrolysis

This is the process whereby electrical energy is used to produc

chemical reaction which will not occur spontaneously. The process co

of passing a direct current through a solution of an electrolyte bet

two electrodes. It is at the electrodes where the chemical changes t place and where the type of process is determined. A typical

electrolysis cell is shown below for the electrolysis of $CuSO_4$

using copper electrodes. Notice that electroneutrality is maintained

by the electrolyte $CuSO_4$ which also carries the current. In this

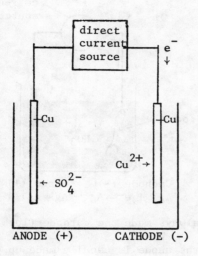

case at the cathode:

$$Cu^{2+}(aq) + 2e^- \rightarrow Cu(s)$$

and Cu plates out on the electrode. At the anode we have

$$Cu(s) \rightarrow Cu^{2+}(aq) + 2e^-$$

and copper is lost from the electrode. The reactions which occur

in electrolytic cells depend on the potentials of the electrolytes

involved and the overall cell construction. If there is more than

one possible reaction, the reaction with the most favourable

potential will occur.

225

Another example of electrolysis is the plating of cutlery or other items (automobile bumper, tools etc.). Plating can be carri out with many metals — Cu, Ni, Cr, Ag, Au. A cell is shown below for the silver plating of a spoon. The electrolyte is $AgNO_3$. The object

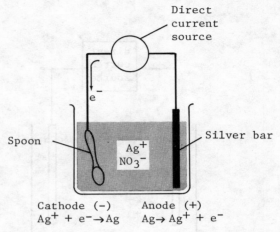

Cathode (−)   Anode (+)
$Ag^+ + e^- \rightarrow Ag$ $Ag \rightarrow Ag^+ + e^-$

to be plated is always made the cathode (look at the half-reaction) while the anode is usually made up of the plating metal (silver in this case). Usually CN^- or other complexing agent is added to the solution to prevent the silver from plating too rapidly (this normally gives an inferior plate). When CN^- is in the solution the electrode reactions are:

ANODE $Ag(s) + 2CN^- \longrightarrow Ag(CN)_2^- + e^-$

        plating metal
        electrode

CATHODE $Ag(CN)_2^- + e^- \longrightarrow Ag(s) + 2CN^-$
(spoon)   (plate)

For another example see Test 15-3. By use of Faraday's Law one can relate the amount of chemical change in an electrolysis

experiment to the amount of electricity used (Section 15-6). At
the end of this section there are some "blank" electrolytic cells
for your use when doing the problems.

5.   The Nernst Equation, Free Energy, and Cell Voltage.

Memorize the following equations:

(a) $\Delta G = -nFE$ where n is the number of moles of

electrons transferred

F is the value of the Faraday in cal $volt^{-1}$

(23,060 cal $volt^{-1}$).

E is the voltage.

At standard conditions, $\boxed{\Delta G° = -23,060 \ nE°}$

(b) Nernst equation.

$$E = E° - \frac{2.303RT \ \log Q}{nF}$$

$$E = E° - \frac{0.059}{n} \log Q \qquad \text{at 25°C.}$$

where E° is the standard potential

Q is the initial quotient.

The Nernst equation relates the cell potential to the concentrations of the species involved. Thus it can be used to determine concentrations of species in solution.

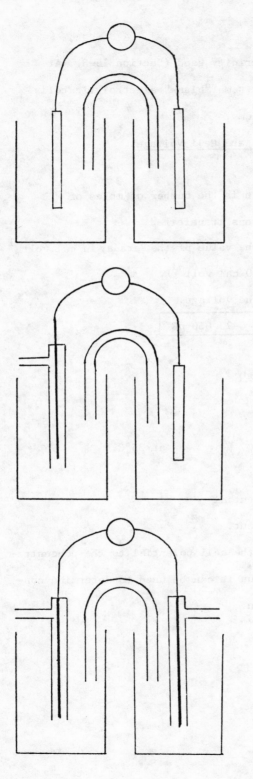

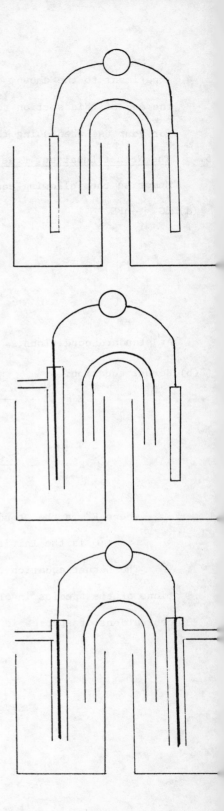

228

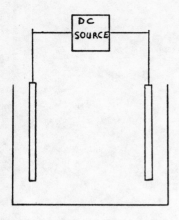 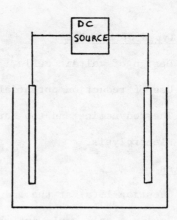

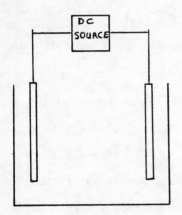 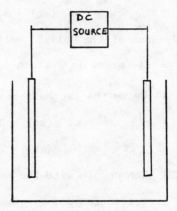

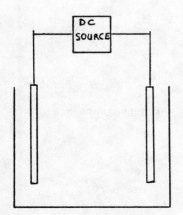 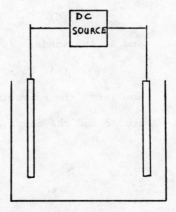

229

B.   Types of Problems

1.   Design of voltaic cells.

2.   Use of reduction potentials.

3.   Thermodynamics and the Nernst equation.

4.   Electrolysis.

Examples

1.(i) Question 19(a) at the end of the chapter.

Design a cell employing the following reaction to generate electric
current:

$$Cu^{2+}(aq) + Ni(s) \longrightarrow Cu(s) + Ni^{2+}(aq)$$

Use one of the "blank" voltaic cells provided and apply the following
rules: (given in Section A,2).

(1)   Write the two half-reactions and indicate their potentials:

$$Cu^{2+}(aq) + 2e^- \longrightarrow Cu(s) \qquad E° = + 0.34 \text{ v}$$

$$Ni^{2+}(aq) + 2e^- \longrightarrow Ni(s) \qquad E° = + 0.250 \text{ v}$$

(2)   The couple with the more positive potential occurs as written
       (reduction ∴ at the cathode) and the other couple occurs in
       reverse (oxidation ∴ at the anode).

       Therefore at the cathode we have:

$$(C) \quad Cu^{2+}(aq) + 2e^- \rightarrow Cu(s) \qquad E° = + 0.34 \text{ v}$$

and at the anode:

$$(A) \quad Ni(s) \rightarrow Ni^{2+}(aq) + 2e^- \qquad E° = - 0.250 \text{ v.}$$

Now label the compartments and indicate which reagents are
needed and the signs of the electrodes.

(3) Electrons always travel toward the positive electrode – the cathode.
Put this on the diagram.

(4) The ions that are reduced, the cations, migrate toward the cathode.
Indicate this on the diagram.

(5) The anions migrate toward the anode. Indicate this. The cell
potential is

$$E^\circ = E^\circ_C + E^\circ_A = + 0.34 + (- 0.250)$$
$$= 0.09 \text{ v.}$$

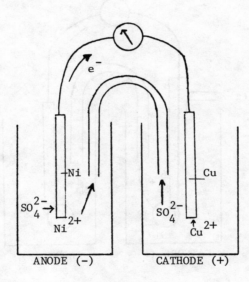

ANODE (−)   CATHODE (+)

(ii) Question 20 at the end of the chapter.

Design a cell employing the following reaction to generate electric
current:

$$Cl_2(g) + 2I^-(aq) \rightarrow 2Cl^-(aq) + I_2(s)$$

231

The half-reactions are:

$$Cl_2(g) + 2e^- \rightarrow 2Cl^-(aq) \qquad E° = + 1.36$$

$$I_2(s) + 2e^- \rightarrow 2I^-(aq) \qquad E° = + 0.535$$

At the cathode we have:

(C) $\qquad Cl_2(g) + 2e^- \rightarrow 2Cl^-(aq)$

At the anode we have:

(A) $\qquad 2I^- \rightarrow I_2(s) + 2e^-$

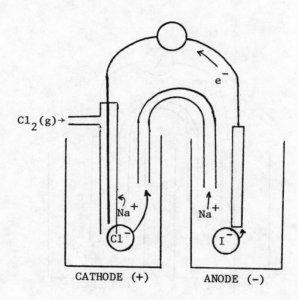

The cell potential is $E° = + 1.36 + (-0.535)$

$$= 0.83 \text{ v.}$$

2.(i) Question 26 at the end of the chapter.

Select an oxidizing agent that will oxidize I^- to I_2 but will not oxidize Cl^- to Cl_2 at standard state conditions.

Write the half reactions:

$$I_2(s) + 2e^- \rightarrow 2I^-(aq) \qquad E° = + 0.535 \text{ v}$$

$$Cl_2(g) + 2e^- \rightarrow 2Cl^-(aq) \qquad E° = + 1.36 \text{ v}$$

Any couple with a reduction potential between +0.535 and +1.36 v will do. Remember I^- (right hand side of half-reaction) will react with the left hand side of any couple below it in Table 15-3. Therefore Fe^{3+}, Ag^+, Hg^{2+}, NO_3^-, Br_2 or O_2 will oxidize I^- but not Cl^-. To oxidize Cl^- we need a couple below 1.36 v in the Table.

(ii) Question 27 at the end of the chapter.

Generate an ordering of half-reactions similar to that in Table 15-3 from the following facts:

(1) $U^{3+} + Cr^{3+} \rightarrow Cr^{2+} + U^{4+}$

(2) $Fe + Sn^{2+} \rightarrow Sn + Fe^{2+}$

(3) $Fe + U^{4+} \rightarrow$ no reaction

(4) $Cr^{2+} + Fe^{2+} \rightarrow$ no reaction

(5) $2Cr^{2+} + Sn^{2+} \rightarrow Sn + 2 Cr^{3+}$

Remember when arranged according to increasing reduction potential, the best reducing agent is in the upper right. When the reduced form of an element is combined with the oxidized form of an element below it in a table of reduction potentials, spontaneous reaction will occur, i.e. the E° for the cell is positive.

From (1) $\quad U^{4+} + e^- \rightarrow U^{3+}$ Since reaction occurs the U^{4+}/U^{3+}

$\quad\quad\quad Cr^{3+} + e^- \rightarrow Cr^{2+}$ potential is more negative than the Cr^{3+}/Cr^{2+} potential.

Therefore $U^{4+} + e^- \rightarrow U^{3+}$ is above $Cr^{3+} + e^- \rightarrow Cr^{2+}$.

From (2) $Fe^{2+} + 2e^- \rightarrow Fe$ Since reaction occurs the Fe^{2+}/Fe

$Sn^{2+} + 2e^- \rightarrow Sn$ couple is more negative than the

Sn^{2+}/Sn couple.

Therefore $Fe^{2+} + 2e^- \rightarrow Fe$ is above $Sn^{2+} + 2e^- \rightarrow Sn$.

From (3) $Fe + U^{4+} \rightarrow$ no reaction.

This means that for the couples

$Fe^{2+} + 2e^- \rightarrow Fe$

$U^{4+} + e \rightarrow U^{3+}$

the Fe^{2+}/Fe potential must be more positive than the U^{4+}/U^{3+}

potential and therefore the $Fe^{2+} + 2e^- \rightarrow Fe$ half-reaction lies

below the $U^{4+} + e \rightarrow U^{3+}$ half-reaction.

From (4) $Cr^{2+} + Fe^{2+} \rightarrow$ no reaction.

This means that for the couples

$$Cr^{3+} + e^- \rightarrow Cr^{2+}$$

$$Fe^{2+} + 2e^- \rightarrow Fe$$

the Cr^{3+}/Cr^{2+} potential must be more positive than the Fe^{2+}/Fe

potential and therefore the $Cr^{3+} + e^- \rightarrow Cr^{2+}$ half-reaction

lies below the $Fe^{2+} + 2e^- \rightarrow Fe$ half-reaction.

From (5) $Cr^{3+} + e^- \rightarrow Cr^{2+}$ Since reaction occurs the Cr^{3+}/Cr^{2+}

$Sn^{2+} + 2e^- \rightarrow Sn$ couple is more negative than the

Sn^{2+}/Sn couple.

Therefore $Cr^{3+} + e^- \rightarrow Cr^{2+}$ is above the $Sn^{2+} + 2e^- \rightarrow Sn$ half-reaction.

234

With the information obtained we can order the half-reactions as below:

$$U^{4+} + e^- \rightarrow U^{3+}$$
$$Fe^{2+} + 2e^- \rightarrow Fe$$
$$Cr^{3+} + e^- \rightarrow Cr^{2+}$$
$$Sn^{2+} + 2e^- \rightarrow Sn$$

increasing

positive

$E°$.

To verify our conclusions we can look up the standard reduction

potentials for the above half-reactions (from top to bottom: -0.61 v,

-0.44 v, -0.41 v, -0.14 v).

(iii) Question 28 at the end of the chapter.

Determine whether or not a redox reaction will occur when you mix the

following at standard state conditions:

(a) H^+, Sn, MnO_4^-

(b) Ag^+, Hg

(c) H^+, Sn

(d) H^+, Cu

(e) Br^-, Pb.

In all of these questions you need to write the reduction potentials

from Table 15-3.

When the reduced form of an element (first couple) is combined with

the oxidized form of an element (second couple) a redox reaction occurs

if the first couple is more negative than the second couple. i.e. $E_1° < E_2°$.

This means that the cell potential must be positive. Therefore

always determine the potential of the cell reaction by adding the half-

reactions. If the potential is positive reaction will occur. If the

potential is negative reaction will not occur.

(a) $\qquad$ $Sn^{2+}(aq) + 2e^- \rightarrow Sn(s)$ $\qquad\qquad\qquad$ $E_1^\circ = -0.14$ v

$\qquad$ $MnO_4^-(aq) + 8H^+(aq) + 5e^- \rightarrow Mn^{2+}(aq) + 4H_2O(\ell)$ $\qquad$ $E_2^\circ = +1.51$ v.

Since $E_1^\circ < E_2^\circ$ or the first couple is above the second couple in Table 15-3 a redox reaction occurs between Sn(s) (reduced form of top couple) and $MnO_4^-(aq)$ (oxidized form of lower couple). It is useful to go through this reasoning for practice. This can also be shown by determining the potential of the required cell:

$\qquad$ 5 x [$\qquad$ $Sn(s) \rightarrow Sn^{2+}(aq) + 2e^-$ $\qquad$] $\qquad\qquad$ $E_1^\circ = +0.14$ v

$\qquad$ 2 x [$MnO_4^-(aq) + 8H^+(aq) + 5e^- \rightarrow Mn^{2+}(aq) + 4H_2O(\ell)$] $\quad$ $E_1^\circ = +1.51$ v

$\qquad$ $\underline{5Sn(s)} + \underline{2MnO_4^-}(aq) + \underline{16H^+}(aq) \rightarrow 5Sn^{2+}(aq) + 2Mn^{2+}(aq)$ $\quad$ $E^\circ = +1.65$ v

$\qquad\qquad\qquad\qquad\qquad\qquad\qquad\qquad\qquad\qquad\qquad$ $+ 8H_2O$ (ℓ)

<u>Note</u> that the half-reactions were multiplied by 5 and by 2 but the standard potentials were not. Since E° is positive a redox reaction will occur.

(b) $\qquad$ $Hg^{2+}(aq) + 2e^- \rightarrow Hg(\ell)$ $\qquad\qquad\qquad$ $E_1^\circ = +0.85$ v

$\qquad$ $Ag^+(aq) + e^- \rightarrow Ag(s)$ $\qquad\qquad\qquad\qquad$ $E_2^\circ = +0.80$ v

and

$\qquad$ $Hg(\ell) \rightarrow Hg^{2+}(aq) + 2e^-$ $\qquad\qquad\qquad$ $E_1^\circ = -0.85$ v

$\qquad$ $\underline{2Ag^+(aq) + 2e^- \rightarrow 2Ag(s)}$ $\qquad\qquad\qquad$ $\underline{E_2^\circ = +0.80}$ v

$\qquad$ $Hg(\ell) + 2Ag^+(aq) \rightarrow Hg^{2+}(aq) + 2Ag(s)$ $\qquad$ $E^\circ = -0.05$ v

Since E° is negative the reaction <u>does not</u> occur.

(c) $\qquad$ $Sn^{2+}(aq) + 2e^- \rightarrow Sn(s)$ $\qquad\qquad\qquad$ $E_1^\circ = -0.14$ v

$\qquad$ $2H^+(aq) + 2e^- \rightarrow H_2(g)$ $\qquad\qquad\qquad$ $E_2^\circ = 0.00$ v

and

$\qquad$ $Sn(s) + 2H^+(aq) \rightarrow Sn^{2+}(aq) + H_2(g)$

236

$$E° = -E_1° + E_2° = + 0.14 \text{ v}$$

Since E° is positive a reaction will occur.

(d) $\qquad$ $2H^+(aq) + 2e^- \rightarrow H_2(g)$ $\qquad\qquad$ $E_1° = 0.00 \text{ v}$

$\qquad\qquad$ $Cu^{2+}(aq) + 2e^- \rightarrow Cu(s)$ $\qquad\qquad$ $E_2° = + 0.34 \text{ v}$

and

$\qquad\qquad$ $2H^+(aq) + 2e^- \rightarrow H_2(g)$ $\qquad\qquad$ $E_1° = 0.00 \text{ v}$

$\qquad\qquad$ $\underline{Cu(s) \rightarrow Cu^{2+}(aq) + 2e^-}$ $\qquad\qquad$ $\underline{E_2° = - 0.34 \text{ v}}$

$\qquad\qquad$ $Cu(s) + 2H^+(aq) \rightarrow Cu^{2+}(aq) + H_2(g)$ $\qquad$ $E° = - 0.34 \text{ v}$

Since E° is negative reaction does not occur.

(e) $\qquad$ $Pb^{2+}(aq) + 2e^- \rightarrow Pb(s)$ $\qquad\qquad$ $E_1° = - 0.13 \text{ v}$

$\qquad\qquad$ $Br_2(\ell) + 2e^- \rightarrow 2Br^-(aq)$ $\qquad\qquad$ $E_2° = 1.087 \text{ v}$

and $\qquad$ $Pb(s)$ and $2Br^-(aq)$ cannot be combined to give an

$\qquad\qquad$ overall reaction. Therefore no reaction occurs.

3.(i) Question 47. at the end of the chapter.

Using the data in Table 15-3, calculate the free energy $\Delta G°$

and K for the reaction

$\qquad\qquad$ $AgI(s) + Zn(s) \rightarrow Zn^{2+}(aq) + Ag(s) + I^-(aq).$

The half-reactions are

$\qquad\qquad$ $AgI(s) + e^- \rightarrow Ag(s) + I^-(aq)$ $\qquad\qquad$ $E° = - 0.15 \text{ v}$

$\qquad\qquad$ $Zn^{2+}(aq) + 2e^- \rightarrow Zn(s)$ $\qquad\qquad$ $E° = - 0.76 \text{ v}$

and $\qquad$ $2AgI(s) + 2e^- \rightarrow 2Ag(s) + 2I^-(aq)$ $\qquad$ $E° = - 0.15 \text{ v}$

$\qquad\qquad$ $\underline{Zn(s) \rightarrow Zn^{2+}(aq) + 2e^-}$ $\qquad\qquad$ $\underline{E° = + 0.76 \text{ v}}$

add: $2AgI(s) + Zn(s) \rightarrow 2Ag(s) + 2I^-(aq) + Zn^{2+}(aq)$ $\quad$ $E° = + 0.61 \text{ v}.$

At standard conditions, $\Delta G° = -23{,}060 \, n \, E°$

$$= -23{,}060 \times 2 \times 0.61$$

$$= -28{,}133 \text{ cal mole}^{-1}$$

$$= -28{,}000 \text{ cal mole}^{-1}. \text{ (significant figures).}$$

But $\Delta G° = -RT\ell nK$

$$= -RT \, 2.303 \, \log K$$

$$\log K = \frac{-\Delta G°}{RT2.303} = \frac{-28{,}000 \text{ cal mole}^{-1}}{1.987 \text{ cal mole}^{-1}\text{deg}^{-1} \times 298 \text{ deg} \times 2.303}$$

$$\log K = 21$$

$$K = 1 \times 10^{21}.$$

(ii) Question 34 at the end of the chapter.

Calculate the voltage produced from the reaction

$$Cd(s) + Co^{2+}(0.10M) \rightarrow Cd^{2+}(1.0M) + Co(s)$$

The half-reactions are

$$Cd^{2+}(aq) + 2e^- \rightarrow Cd(s) \qquad\qquad E° = -0.40 \text{ v}$$

$$Co^{2+}(aq) + 2e^- \rightarrow Co(s) \qquad\qquad E° = -0.28 \text{ v}$$

and for the reaction

$$Cd(s) \rightarrow Cd^{2+}(aq) + 2e^- \qquad\qquad E° = +0.40 \text{ v}$$

$$\underline{Co^{2+}(aq) + 2e^- \rightarrow Co(s) \qquad\qquad E° - 0.28 \text{ v}}$$

$$Cd(s) + Co^{2+}(aq) \rightarrow Cd^{2+}(aq) + Co(s) \qquad E° = +0.12 \text{ v}$$

The Nernst equation is

$$E = E° - \frac{0.059}{n} \log \frac{[Cd^{2+}]}{[Co^{2+}]}$$

$$= +0.12 - \frac{0.059}{2} \log 10$$

$$= 0.12 - 0.03$$

$$= 0.09 \text{ v.}$$

(iii) Question 35 at the end of the chapter.

A cell is constructed based on the reaction

$$Zn(s) + 2H^+(aq) \rightarrow Zn^{2+}(1.0M) + H_2 \text{ (1atm).}$$

The potential is measured at 0.50 v. Calculate $[H^+]$.

$$E = E° - \frac{0.059}{n} \log \frac{[Zn^{2+}]P_{H_2}}{[H^+]^2}$$

The two half-reactions are

$$Zn^{2+}(aq) + 2e^- \rightarrow Zn(s) \qquad\qquad E° = -0.76 \text{ v}$$

$$2H^+(aq) + 2e^- \rightarrow H_2(g) \qquad\qquad E° = 0.00$$

and the reaction is

$$Zn(s) \rightarrow Zn^{2+}(aq) + 2e^- \qquad\qquad E° = + 0.76 \text{ v}$$

$$\underline{2H^+(aq) + 2e^- \rightarrow H_2(g) \qquad\qquad\qquad E° = 0.00 \text{ v}}$$

$$Zn(s) + 2H^+(aq) \rightarrow Zn^{2+}(aq) + H_2(g) \qquad E° = + 0.76 \text{ v}$$

$$\therefore \quad 0.50 = 0.76 - \frac{0.059}{2} \log \frac{[1.0] \times 1.0}{[H^+]^2}$$

$$\frac{0.059}{2} \log \frac{[1.0] \times 1.0}{[H^+]^2} = 0.26$$

$$\log \frac{1}{[H^+]^2} = \frac{2 \times 0.26}{0.059}$$

$$\log [H^+]^2 = -8.8$$

$$= -9 + 0.20$$

$$[H^+]^2 = \text{antilog} (-9) \times \text{antilog} (0.20)$$

$$= 1.6 \times 10^{-9}$$
$$[H^+] = 4.0 \times 10^{-5}.$$

4. Question 40 at the end of the chapter.

In Exercise 18(c) you designed the Edison cell. Draw the electrolysis cell and indicate the reactions taking place during charging.

Recharging any battery simply uses an external source of electricity to drive the electrons through the battery in the opposite direction to that for the voltaic cell. The chemical changes are reversed.

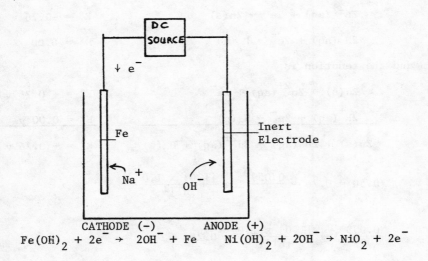

CATHODE (−)   ANODE (+)

$$Fe(OH)_2 + 2e^- \rightarrow 2OH^- + Fe \qquad Ni(OH)_2 + 2OH^- \rightarrow NiO_2 + 2e^-$$

C.   Test Yourself

1.   Design a cell using the following reaction to generate electric
      current: $Cd(s) + Ni^{2+}(aq) \rightarrow Cd^{2+}(aq) + Ni(s)$

2.   Design a cell using the following reaction to generate electric
      current: $H_2(g) + Cl_2(g) \rightarrow 2H^+(aq) + 2Cl^-(aq)$

3.   Design a cell using the following reaction to generate electric
      current: $Fe(s) + O_2(g) + 4H^+(aq) \rightarrow Fe^{2+}(aq) + 2H_2O(\ell)$

4.(a) Design a cell using the following two half-reactions to generate
      electric current.

$$Zn(OH)_2 + 2e^- \rightarrow Zn + 2OH^- \qquad\qquad E° = -1.25 \text{ v}$$
$$2AgO + H_2O + 2e^- \rightarrow Ag_2O + 2OH^- \qquad\qquad E° = +0.57 \text{ v.}$$

   (b) Calculate the cell potential.

   (c) Draw the cell when it is being recharged and indicate the reactions
       that occur at the anode and cathode.

5.   Determine whether or not a useful redox reaction will occur when
      you mix the following at standard state conditions:

   (a) $Mn(s)$, $H^+(aq)$

   (b) $Ag^+(aq)$, $Br^-(aq)$

   (c) $MnO_4^-(aq)$, $H^+(aq)$, $NO(g)$

   (d) $Fe^{3+}(aq)$, $OH^-(aq)$, $H_2(g)$.

6.   Using the data in Table 15-3, calculate $\Delta G°$ and K for the
      reaction

$$Cd(s) + Cl_2(g) \rightarrow Cd^{2+}(aq) + 2Cl^-(aq).$$

241

7.   For the cell reaction

$$Cd(s) + 2H^+(aq) \rightarrow Cd^{2+} (1.0M) + H_2 (1 \text{ atm}).$$

The potential is measured at 0.25 V. Calculate $[H^+]$.

8.   Draw a cell for the electrolysis of $Cu(SO_4)$

   (a) between two inert electrodes and

   (b) between two copper electrodes.

   Write all equations and justify them.

16

The Interaction of Matter with Electromagnetic Radiation

A. <u>Points of Importance</u>

1. Review chapter 4 and the equation

$$E = \frac{hc}{\lambda} = h\nu$$

Remember <u>high energy</u>, <u>high frequency</u>, and <u>short wavelength</u> go together just as <u>low energy</u>, <u>low frequency</u>, and <u>long wavelength</u>.

2. <u>Beer's Law</u>

$$A = C \times \varepsilon \times \ell$$

where A = absorbance of a sample

C = concentration in M of the absorbing species

ε = molar absorptivity or extinction coefficient
for the species

ℓ = path length of the cell in cm.

For a 1 cm path cell and a species R we have

$$A = [R] \times \varepsilon_R \text{ which is equation 16-2 in the text.}$$

If there are two species R and S that absorb at the same wavelength then the total absorbance of the solution is given by

$$A_T = [R] \times \varepsilon_R + [S] \times \varepsilon_S \quad \text{(for a 1 cm cell).}$$

3. New Terms

singlet state: a state with no unpaired electrons.

triplet state: a state with two unpaired electrons.

quantum yield: number of moles reacted per number of moles of photons
absorbed.

monochromatic radiation: light consisting of a single wavelength.

phosphorescence: when an excited state returns to the ground state by
a triplet to singlet transition.

fluorescence: when an excited state returns to the ground state by a singlet
to singlet transition.

einstein: one mole of photons.

244

B.   Types of Problems

1.   Calculation of wavelength, frequency, and energy.

2.   Use of Beer's Law.

Examples

1.(i) Question 5. at the end of the chapter.

Given a wave number of 8000. cm^{-1} calculate the

(a)   wavelength

(b)   frequency

(c)   energy.

(a) $\lambda = \frac{1}{\bar{\nu}} = \frac{1}{8000. \, cm^{-1}} = 1.250 \times 10^{-4}$ cm.

(b) $\nu = c\bar{\nu} = 3.000 \times 10^{10} cm \, s^{-1} \times 8000. \, cm^{-1}$

$= 2.400 \times 10^{14} \, s^{-1}$

(c) $E = hc\bar{\nu} = h\nu = 6.6 \times 10^{-27} erg\text{-}sec \times 2.400 \times 10^{14} \, s^{-1}$

$= 1.6 \times 10^{-12}$ erg.

(ii)Question 11. at the end of the chapter.

Which is the highest energy radiation: visible, infrared, or

microwave?

Since $E = \frac{hc}{\lambda}$, the energy is proportional to $\frac{1}{\lambda}$. The wavelengths

of the three types of radiation are given below:

visible $\lambda \simeq 10^{-5} cm$

infrared $\lambda \simeq 10^{-3} cm$

microwave $\lambda \simeq 1$ cm

Therefore visible light is the highest energy radiation.

2.   Question 14 at the end of the chapter.

A solution of $I_2 (\varepsilon = 925 \, M^{-1} cm^{-1}$ at 516 nm) and a complex $B:I_2$

(ε = 800 M^{-1}cm^{-1} at 516 nm) is studied, where B is a Lewis base that does not absorb at this wavelength. The base concentration at equilibrium is 3.00 x 10^{-4} M and the sum of the concentrations of I$_2$ and BI$_2$ is 1.50 x 10^{-4} M. The equilibrium constant for the reaction

$$B + I_2 \rightleftarrows B:I_2$$

is 3.80. Calculate the absorbance of the solution at 516 nm in a 1 cm cell.

In this problem we need to calculate the concentrations of I$_2$ and BI$_2$ in order to use ε_{I_2} and ε_{BI_2} to obtain the <u>total absorbance</u> of the solution

$$c_B = 3.00 \times 10^{-4}$$

$$c_{I_2} + c_{BI_2} = 1.50 \times 10^{-4} \text{ M}$$

$$\therefore c_{BI_2} = 1.50 \times 10^{-4} \text{ M} - c_{I_2}$$

Substitute these values into the equilibrium expression and solve:

$$K = 3.80 = \frac{[BI_2]}{[I_2]\,[B]}$$

$$3.80 = \frac{1.50 \times 10^{-4} - c_{I_2}}{c_{I_2} \times 3.00 \times 10^{-4}}$$

Solving we obtain c_{I_2} = 1.50 x 10^{-4} M.

$$\therefore \text{ we conclude } c_{BI_2} = 0.$$

$$A \text{ total} = \varepsilon_{I_2} \times c_{I_2} + \varepsilon_{BI_2} \times c_{BI_2}$$

$$= 925 \times 1.50 \times 10^{-4}$$

$$= 0.139.$$

C. Test Yourself

1. Which has the higher energy: a microwave with a frequency of 4.0×10^9 or an ultraviolet ray with a wavelength of 1×10^{-6} cm?

2. Given a frequency of $3.0 \times 10^6 s^{-1}$ calculate

    (a) energy

    (b) wavelength

3. Calculate the energy of the following types of radiation:

    (a) X-rays, $\lambda = 2.0 \times 10^{-8}$ cm.

    (b) Ultraviolet, $\lambda = 1.0 \times 10^{-6}$ cm.

    (c) Infrared, $\lambda = 1.5 \times 10^{-4}$ cm.

    (d) Radio, $\nu = 2.0 \times 10^8 s^{-1}$

4. Solutions of chromate ion, CrO_4^{2-}, in basic solution are yellow. At 372 nm chromate has a molar absorptivity of 4815 $M^{-1}cm^{-1}$. Calculate the concentration of chromate if the absorbance is 0.750 in a 2 cm cell.

5. A solution of an organic dye gave an absorbance of 0.942 in a 5- cm cell at 600 nm. The solution was 5.80×10^{-4} M in the dye. Calculate the molar absorptivity at 600 nm.

17

Nuclear Sources of Energy

A. Points of Importance

1. New Terms

Nuclear fission: the process which occurs when a nucleus captures

neutrons and splits into two new nuclei of roughly

equal size.

Nuclear fusion: the process which occurs when the nuclei of two light

atoms are fused together to form one nucleus.

Alpha ray: a helium nucleus, $_2^4 \text{He}$.

Beta ray: an electron, $_{-1}^0 \text{e}^-$.

Gamma ray: short wavelength electromagnetic radiation.

Binding energy: the energy required to convert a nucleus to its

constituent nuclear particles (nucleons). Equals

the mass-defect.

249

Half-life: the time required for half the original sample to
               disappear.

Mass-defect: the mass difference between actual nuclei and the
               sums of the masses of their constituent nuclear
               particles (nucleons).

2. Balancing nuclear equations

The sum of the mass numbers (and atomic numbers) on the left-hand side
of the equation must equal the sum of those on the right. Note for the
equation:

$$^{230}_{90}\text{Th} \longrightarrow {}^{226}_{88}\text{Ra} + {}^{4}_{2}\text{He}$$

mass number left-hand side   = 230

atomic number left-hand side = 90

mass number right-hand side   = 226 + 4 = 230

atomic number right-hand side = 88 + 2 = 90.

3. Binding Energy

The binding energy can be calculated by determining the mass defect as
in Test 17-2. The binding energy per nucleon gives a measure of the
stability of a nucleus toward radioactive decay. The binding energy per
nucleon is given below for a number of nuclei.

Species	Binding Energy/nucleon (MeV).
$^{4}_{2}\text{He}$	7.08
$^{12}_{6}\text{C}$	7.67
$^{35}_{17}\text{Cl}$	8.52

Species	Binding Energy/nucleon (MeV).
$^{133}_{55}Cs$	8.41
$^{235}_{92}U$	7.59

The lower value for the heavy nuclide, $^{235}_{92}U$, means that when $^{235}_{92}U$ undergoes fission, the mass defect in the smaller particles (products) is greater, so some mass has been converted into energy. i.e.

$$^{235}_{92}U + ^{1}_{0}n \longrightarrow ^{141}_{56}Ba + ^{92}_{36}Kr + 3^{1}_{0}n \left.\begin{array}{l} \text{} \\ \text{} \\ \text{} \end{array}\right\} \begin{array}{l} \text{products are} \\ \text{more stable and} \\ \text{energy is released.} \end{array}$$

Binding Energy 7.59   ~8.4   ~8.8
per nucleon
  (MeV)

One can also calculate the energy change of a nuclear reaction using Einstein's equation

$$\Delta E = \Delta mc^{2}.$$

We require the change in mass, Δm, defined as

$$\Delta m = \text{mass of products} - \text{mass of reactants.}$$

4. Some of the common particles and their symbols are given on the next page for handy reference.

Particle	Mass(amu)	Charge	Symbol	Comment
neutron	1.00867	0	$_{0}^{1}n$	
proton	1.00728	+1	$_{1}^{1}H$	
alpha	4.00150	+2	$_{2}^{4}He$	
beta	0.000549	−1	$_{-1}^{0}e^{-}$	electron
gamma	0		γ	electromagnetic radiation
positron	0.000549	+1	$_{1}^{0}e^{-}$	"positive electron"

B.   Types of Problems

1.   Balancing nuclear equations.

2.   Calculation of mass defect and binding energy.

Examples

1.   Questions 5.(a), (c), (d) and (e) at the end of the chapter.

      Write balanced equations for the reaction that occurs when

      (a) $^{239}_{93}$Np emits a beta particle.

      (c) $^{210}_{81}$Tℓ emits an alpha particle.

      (d) $^{203}_{81}$Pb captures an electron and emits a gamma ray

      (e) $^{110}_{49}$In emits a positron.

   (a) $^{239}_{93}$Np $\longrightarrow$ $^{0}_{-1}e^{-}$ + $^{239}_{94}$Pu

   (c) $^{210}_{81}$Tℓ $\longrightarrow$ $^{4}_{2}$He + $^{206}_{79}$Au

   (d) $^{203}_{81}$Pb + $^{0}_{-1}e$ $\longrightarrow$ $^{203}_{80}$Hg + gamma

   (e) $^{110}_{49}$In $\longrightarrow$ $^{0}_{1}e^{-}$ + $^{110}_{48}$Cd

2.(i) Question 11 at the end of the chapter.

      Calculate the binding energy per nucleon for $^{56}_{26}$Fe, whose nuclear

      mass is 55.9207 amu.

      Mass of 26 protons = 26 (1.00728) = **26.18928 amu**

      Mass of 30 neutrons = 30 (1.00867) = 30.26010 amu

      56.44938 amu

      Mass defect = mass of all nuclear particles − actual mass

                  = 56.44938 − 55.9207

253

$$= 0.5287 \text{ amu}$$

$$\text{Binding energy} = 0.5287 \text{ amu} \times 931.5 \frac{\text{MeV}}{\text{amu}} = 492.51 \text{ MeV}$$

$$\text{Binding energy/nucleon} = \frac{492.51 \text{ MeV}}{56 \text{ nucleons}} = 8.795 \frac{\text{MeV}}{\text{nucleon}}.$$

(ii) Question 14 at the end of the chapter.

The mass defect for $^{196}_{80}\text{Hg}$ is 1.6653 amu. Calculate the binding

energy per nucleon.

$$\text{Binding energy} = 1.6653 \text{ amu} \times 931.5 \frac{\text{MeV}}{\text{amu}} = 1551 \text{ MeV}$$

$$\text{Binding energy per nucleon} = \frac{1551 \text{ MeV}}{196 \text{ nucleons}} = 7.913 \frac{\text{MeV}}{\text{nucleon}}.$$

(iii) Question 16 at the end of the chapter.

Calculate ΔE per atom for the reaction

$$^{1}_{0}\text{n} + ^{40}_{19}\text{K} \longrightarrow ^{37}_{17}\text{Cl} + ^{4}_{2}\text{He}$$

given that the mass of ^{40}K is 39.95358 amu, the mass of ^{37}Cl is

36.95657 amu and the mass of helium is 4.00150 amu.

$$\Delta m = \text{mass of products} - \text{mass of reactants}$$

$$= [4.00150 + 36.95657] - [39.95358 + 1.00867]$$

$$= 40.95807 - 40.96225$$

$$= -0.00418 \text{ amu}$$

$$\Delta E = -0.00418 \text{ amu} \times 931.5 \frac{\text{MeV}}{\text{amu}} = -3.89 \text{ MeV}$$

(iv) Question 17 at the end of the chapter.

Carry out calculations to determine whether or not the following

reaction would be exothermic:

$$^9_4\text{Be} + ^{10}_5\text{B} \longrightarrow ^{19}_9\text{F}.$$

The nuclear masses of the species are:

$^9_4\text{Be} = 9.0099\text{amu}, \quad ^{10}_5\text{B} = 10.01019 \text{ amu}, \quad ^{19}_9\text{F} = 18.99346 \text{ amu}.$

$\Delta m = 18.99346 - [9.0099 + 10.01019]$

$\quad = 18.99346 - 19.02018$

$\quad = -0.02672 \text{ amu}$

$\Delta E = -0.02672 \text{ amu} \times 931.5 \dfrac{\text{MeV}}{\text{amu}} = -24.89 \dfrac{\text{MeV}}{\text{amu}}.$

The reaction is exothermic.

C. Test Yourself

1. Write balanced equations for the reaction that occurs when

(a) $^{235}_{92}U$ emits a beta particle.

(b) $^{131}_{53}I$ emits a beta particle.

(c) $^{43}_{21}Sc$ emits a proton.

(d) $^{38}_{19}K$ emits a positron.

(e) $^{197}_{80}Hg$ captures an electron.

(f) $^{238}_{92}U$ emits an alpha particle.

(g) $^{253}_{99}Es$ captures an alpha particle and emits a neutron.

(h) $^{10}_{5}B$ captures a neutron.

2. Fill in the missing species in the following equations.

(a) $^{235}_{92}U + ^{1}_{0}n \longrightarrow ^{148}_{58}Ce +$ _____ $+ 3 ^{1}_{0}n$

(b) $^{235}_{92}U + ^{1}_{0}n \longrightarrow ^{134}_{52}Te +$ _____ $+ 3 ^{1}_{0}n$

(c) $^{235}_{92}U + ^{1}_{0}n \longrightarrow ^{94}_{38}Sr +$ _____ $+ 3 ^{1}_{0}n$

(d) $^{218}_{85}At \longrightarrow$ _____ $+ ^{214}_{83}Bi$

3. Calculate the binding energy per nucleon for $^{18}_{8}O$, whose nuclear mass is 17.99477 amu.

4. Calculate the binding energy per nucleon for $^{146}_{58}Ce$, whose nuclear mass is 145.8865 amu.

5. Calculate ΔE per atom for the reaction

$^{238}_{92}U + ^{1}_{0}n \longrightarrow ^{239}_{94}Pu + 2 ^{0}_{-1}e$. The atomic masses are $^{238}_{92}U$ = 238.0003,

$$^{239}_{94}Pu = 239.0006.$$

6. Calculate ΔE per atom for the reaction

$$^{239}_{94}Pu + ^{1}_{0}n \longrightarrow ^{144}_{58}Ce + ^{90}_{38}Sr + 2\ ^{0}_{-1}e + 6\ ^{1}_{0}n.$$ The atomic masses are

$$^{239}_{94}Pu = 239.0006, \quad ^{144}_{58}Ce = 143.8816, \quad ^{90}_{38}Sr = 89.8864 \text{ amu}.$$

7. Calculate ΔE per atom for the reaction

$$^{2}_{1}H + ^{3}_{1}H \longrightarrow ^{4}_{2}He + ^{1}_{0}n.$$ The atomic masses are:

$$^{2}_{1}H = 2.01355, \quad ^{3}_{1}H = 3.01550, \quad ^{4}_{2}He = 4.00150 \text{ amu}.$$

8. Using only the species given below and neutrons and electrons

    (a) give an example of a fission reaction and calculate ΔE.

    (b) give an example of a fusion reaction and calculate ΔE.

    Species: $\quad ^{235}_{92}U = 234.9934, \quad ^{2}_{1}H = 2.01355, \quad ^{90}_{38}Sr = 89.8864,$

    $$^{4}_{2}He = 4.00150, \quad ^{144}_{58}Ce = 143.8816, \quad ^{1}_{0}n = 1.0087,$$

    $$^{0}_{-1}e = 0.00055 \text{ amu}.$$

18

Chemical Kinetics

A. <u>Points of Importance</u>

1. <u>Rates of reactions</u>

The <u>rate of reaction</u>, r, is defined as the amount of reactant consumed (or product produced) during a given period of time. For the reaction A ———> B the rate of reaction = $\dfrac{\text{change in concentration of A or B}}{\text{elapsed time}}$.

Generally, the <u>rate of a reaction</u> decreases with time because it depends on the concentration of reactants. This means that as A is used up the concentration of A remaining is less and the reaction rate decreases. This can be seen in the diagram below

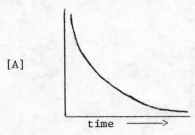

For the reaction A $\longrightarrow$ B this can be expressed mathematically as $r = -\dfrac{dA}{dt} = \dfrac{dB}{dt}$. For the general reaction aA + bB = cC + dD the rate of reaction is expressed as $r = -\dfrac{1}{a}\dfrac{dA}{dt} = -\dfrac{1}{b}\dfrac{dB}{dt} = \dfrac{1}{c}\dfrac{dC}{dt} = \dfrac{1}{d}\dfrac{dD}{dt}$.

As stated above the rate of reaction depends on the concentration of the reactants. The manner in which r depends on the concentration of the reactants can be expressed mathematically and is called the <u>rate law</u>. The <u>rate law</u> is determined <u>experimentally</u>. For example, for A $\longrightarrow$ B

$$r = \text{constant } [A]^n$$

and n is determined experimentally. The constant is called the rate constant and is given the symbol k: $\qquad r = k\,[A]^n$.

$$\text{The units of } r \text{ are } \frac{\text{mole}}{\text{liter sec}}\ .$$

The value of n is the <u>order of the reaction</u> in the reactant A. If we had a more complicated equation

$$A + B \longrightarrow P$$

then

$$r = k\,[A]^n[B]^m$$

In this case the order with respect to [A] is n and with respect to [B] is m. The <u>overall order</u> is n + m. As pointed out in section 18-3 of the text the more usable form of the rate equation is a simple expression relating concentration to time. For the reaction A $\longrightarrow$ B where n = 1 the rate of reaction, $r = -\dfrac{dA}{dt} = k\,[A]$, can be converted by integration to

$$\log\,[A]_t = -\frac{kt}{2.303} + \log\,[A]_i \quad \text{where } [A]_i \text{ is the initial}$$

$$\log\,\frac{[A]_i}{[A]_t} = \frac{kt}{2.303}\ . \qquad \begin{array}{l}\text{concentration and } [A]_t \text{ is the} \\ \text{concentration at time t.}\end{array}$$

Thus k can be obtained from the slope of a plot of $\log\frac{[A]i}{[A]_t}$ vs time. It is much more convenient to determine k, which is a constant, in this manner than it is to have a table of reaction rates at different times. Table 18-1 gives the integrated forms of the rate laws for various orders n.

A useful parameter in chemical kinetics is the half-life, $t_{\frac{1}{2}}$, the time required for one half of the initial amount of material to disappear. The half-life for various order reactions are given below:

order	half-life
1st	$t_{\frac{1}{2}} = \dfrac{0.693}{k}$
2nd	$t_{\frac{1}{2}} = \dfrac{1}{k[A]_i}$
zero	$t_{\frac{1}{2}} = \dfrac{[A]_i}{2k}$

2. Mechanism

Mechanisms are postulated so that they are consistent with the experimentally determined rate law. The rate law refers to the slowest step of all the reactions that lead to the overall reaction and those that precede it. For the reaction

$$2NO_2Cl \longrightarrow 2NO_2 + Cl_2 \quad \text{a possible mechanism is}$$

$$NO_2Cl \xrightarrow{k} NO_2 + Cl \qquad \text{slow step}$$

$$NO_2Cl + Cl \longrightarrow NO_2 + Cl_2 \qquad \text{fast step}$$

The rate law for this reaction is

$$\frac{dCl_2}{dt} = k[NO_2Cl].$$ Remember there are no reactions preceding the slow step.

260

Say you are given the rate law for the reaction A + B + 2C ———> D + E
as rate = k[A][C]. What is a possible mechanism? Since the slow step
and those that precede it determine the rate law a <u>possible</u> mechanism is

$$A + C \longrightarrow D + R \qquad \text{slow}$$

$$R + B \longrightarrow X \qquad \text{fast}$$

$$\underline{X + C \longrightarrow E} \qquad \text{fast}$$

The overall
reaction is   A + B + 2 C ———> D + E.

3. <u>Temperature and Reaction Rates</u>

The variation of the rate constant with temperature is given by the
equation

$$\log k = \log A - \frac{E_a}{2.303RT}$$

E_a is the activation energy for the reaction and log A is the
frequency factor (defined in text).

A reaction path diagram for an endothermic reaction is shown
below:

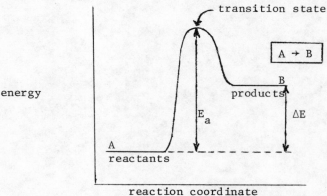

Note that the <u>transition state</u> is the point of highest energy on the
curve. If an intermediate existed for an endothermic reaction the

261

diagram would be

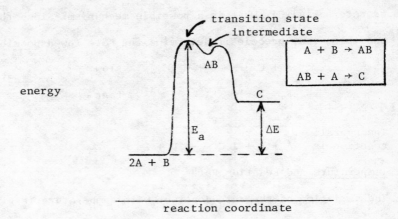

energy

reaction coordinate

B.   Types of Problems

1.   Writing the rate law and determining the order for postulated

      mechanisms.

2.   Determining the rate law from experimental data.

3.   Use and determination of $t_{1/2}$.

4.   Calculation of activation energies.

Examples

1.(i) Question 6. at the end of the chapter.

      In a hypothetical reaction of A with B, it is found that when the [B]

      is doubled with [A] held constant, the initial reaction rate quadruples.

      When the concentration of [B] is held constant and [A] is doubled,

      there is no change in rate. What is the rate law?

          Since the rate quadruples on doubling [B] we have a $[B]^2$ dependence.

      The rate is not dependent on [A] therefore [A] does not occur in the

      rate law.

$$\frac{-d[A]}{dt} = k\ [B]^2 = rate$$

  (ii) Question 7. at the end of the chapter.

      The rate law for a reaction is found to be

$$\frac{d[A]}{dt} = k\ [A]^{1/2}$$

        (a) What is the order?

        (b) What are the units of k?

      (a) The order is $1/2$.

      (b) The units of $\frac{d\ [A]}{dt}$ are $\frac{M}{sec}$

$$\therefore \quad \frac{d\,[A]}{dt} = k\,[A]^{\frac{1}{2}}$$

$$k = \frac{d\,[A]}{dt}\ \frac{1}{[A]^{\frac{1}{2}}} = \frac{M}{sec} \times \frac{1}{M^{\frac{1}{2}}} = M^{\frac{1}{2}}\ sec^{-1}$$

2. Question 19. at the end of the chapter.

Compound A decomposes as follows:

time (sec)	0	10	20	30
[A]	0.500	0.357	0.278	0.227

Determine the order and calculate the rate constant.

In order to determine the order we will try various plots of the data according to the equations in Table 18-1.

First try a zero order plot: $[A]_t = [A]_i - kt$.

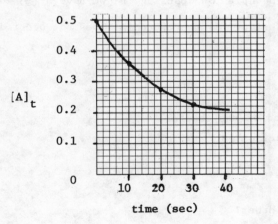

Since the plot is not a straight line the reaction is not zero order.

Now try a first order plot: $\log[A]_t = \frac{-kt}{2.303} + \log[A]_i$

time(sec)	$-\log[A]_t$
0	0.301
10	0.447
20	0.556
30	0.644

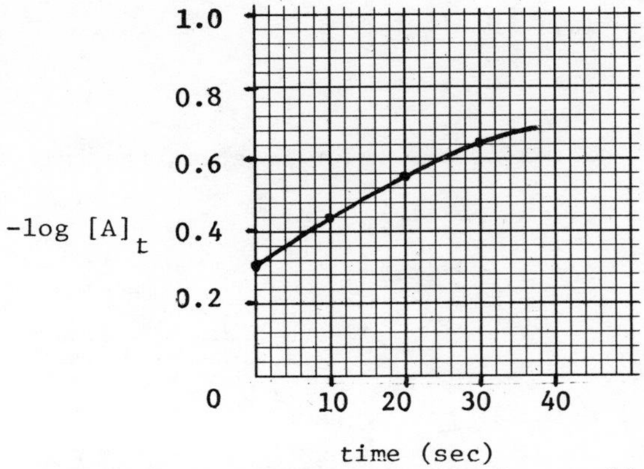

$-\log[A]_t$

time (sec)

The first order plot is not straight and the reaction is not first

order.

Next try a second order plot: $\dfrac{1}{[A]_t} - \dfrac{1}{[A]_i} = kt$

time(sec)	$\dfrac{1}{[A]_t}$
0	2.00
10	2.80
20	3.60
30	4.41

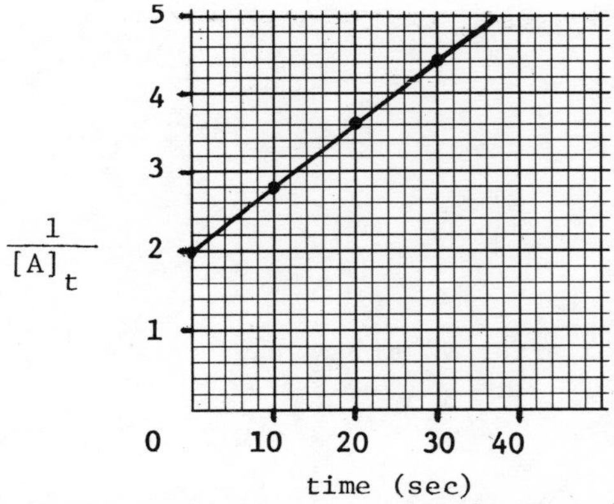

$\dfrac{1}{[A]_t}$

time (sec)

This plot gives a straight line therefore the reaction is second

order. The slope of this line is equal to k.

$$\text{slope} = \frac{(4.41 - 2.00)}{(30 - 0)} \frac{M^{-1}}{\sec} = 0.0803 \ M^{-1}\sec^{-1}.$$

3. Question 14. at the end of the chapter.

The half-life of a 0.15 M solution of A is 30 minutes.

The reaction is second order. How much A remains after

(a) 10 minutes?

(b) 90 minutes?

For a second order reaction the half-time is

$$t_{\frac{1}{2}} = \frac{1}{k[A]_i}$$

$$\therefore \ k = \frac{1}{t_{\frac{1}{2}}[A]_i} = \frac{1}{30 \ min \ x \ \ 0.15 \ M} = 0.22 \ M^{-1}min^{-1}.$$

(a) after 10 min we have

$$\frac{1}{[A]_t} - \frac{1}{[A]_i} = kt$$

$$\frac{1}{[A]_t} = kt + \frac{1}{[A]_i}$$

$$= 0.22 \ M^{-1}min^{-1} \ x \ 10 \ min + \frac{1}{0.15 \ M}$$

$$= 2.2 + 6.7$$

$$= 8.9$$

$$[A]_t = 0.11$$

(b) after 90 min we have

$$\frac{1}{[A]_t} = 0.22 \ M^{-1}min^{-1} \ x \ 90 \ min + \frac{1}{0.15 \ M}$$

$$= 19.8 + 6.7$$

$$= 26.5$$

$$[A]_t = 0.038.$$

4.(i) Question 25. at the end of the chapter.

The rate constant for the gas phase reaction

$$H_2 + I_2 \ \longrightarrow \ 2HI$$

changes with temperature as follows:

T(°K)	556	575	629	666	781
k	4.5×10^{-5}	1.4×10^{-4}	2.5×10^{-3}	1.4×10^{-2}	1.34

Calculate the activation energy.

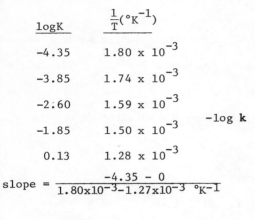

logK	$\frac{1}{T}(°K^{-1})$
-4.35	1.80×10^{-3}
-3.85	1.74×10^{-3}
-2.60	1.59×10^{-3}
-1.85	1.50×10^{-3}
0.13	1.28×10^{-3}

$$\text{slope} = \frac{-4.35 - 0}{1.80 \times 10^{-3} - 1.27 \times 10^{-3} \ °K^{-1}}$$

$$= \frac{-4.35}{0.53 \times 10^{-3} \ °K^{-1}}$$

$$= -8.21 \times 10^{3} \ °K$$

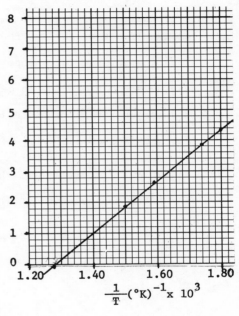

From the equation (18-26):

$$\log k = \log A - \frac{E_a}{2.303 \ RT}$$

$$\therefore \ \text{slope} = - \frac{E_a}{2.303 \ R} = - 8.21 \times 10^{3} \ °K$$

$$E_a = 8.21 \times 10^{3} °K \times 2.303 \times 1.987 \ cal \ mole^{-1} °K^{-1}$$

$$E_a = 37.6 \times 10^{3} \ cal \ mole^{-1}.$$

(ii) Question 27 at the end of the chapter.

In which of the following cases would you expect to have the best chance of isolating the intermediate? Why? Diagrams are shown on p. 268.

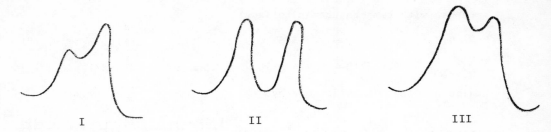

I   II   III

In case II one would have the best chance of isolating an intermediate

since the intermediate, once formed, has stability comparable to the

reactants and products. In cases I and III the intermediate may decay to

products or reactants since it is of high energy with respect to

reactants and products.

(iii) **Question 28 at the end of the chapter.**

(a) Label the quantities (a), (b), and (c) in the following

diagram:

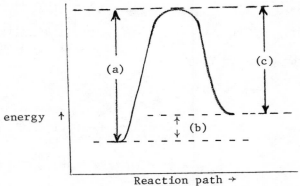

(b) What is the sign of ΔE for the reaction in (a)?

(a) In the diagram (a) is E_a, the activation energy for the forward reaction,

(b) is the internal energy change for the reaction, ΔE,

(c) is E_a^1, the activation energy for the reverse reaction.

268

(iv) Question 29. at the end of the chapter.

The free energy of formation of HI(g) is + 0.3 kcal mole^{-1} and the activation energy (assume to be a free energy) for the reaction $H_2(g) + I_2(g) \longrightarrow 2HI(g)$ is 38.9 kcal mole^{-1}. Calculate the activation energy for the reaction $2HI(g) \longrightarrow H_2(g) + I_2(g)$.

Since the free energy of formation of HI(g) is + 0.3 kcal mole^{-1} the internal energy ΔE is positive and we have an endothermic reaction. This is illustrated below:

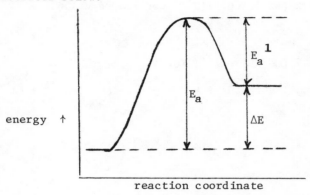

From this diagram we have the following relation

$$E_a = E_a^{\,1} + \Delta E$$

where E_a is the activation energy for the forward reaction,

$E_a^{\,1}$ is the activation energy for the reverse reaction,

ΔE is the internal energy change.

$$\therefore \quad E_a^{\,1} = E_a - \Delta E$$
$$= 38.9 - 0.3$$
$$E_a^{\,1} = 38.6 \text{ kcal mole}^{-1}.$$

C.   Test Yourself.

1.   What are the orders of the reactions in each reactant and the overall orders for the following rate laws?

(a) rate = k $[Fe^{2+}][OH^-]^2[O_2]$ for the reaction

$$4Fe^{2+} + O_2 + 2H_2O \longrightarrow 4Fe^{3+} + 4OH^-.$$

(b) rate = k $[NO]^2[O_2]$ for the reaction

$$2NO + O_2 \rightarrow 2NO_2.$$

(c) rate = k $[Br^-][BrO_3^-][H^+]^2$ for the reaction

$$5Br^- + BrO_3^- + 6H^+ \longrightarrow 3Br_2 + 3H_2O.$$

(d) rate = k $[Co(NH_3)_5Cl^{2+}]$ for the reaction

$$Co(NH_3)_5Cl^{2+} + H_2O \longrightarrow Co(NH_3)_5OH_2^{3+} + Cl^-.$$

2.   Data for the reaction

$$A + B \xrightarrow{\ k\ } \text{products}$$

are given below:

[A]	[B]	rate(Msec^{-1})
0.025	0.025	0.58
0.025	0.050	1.16
0.025	0.100	2.32
0.050	0.100	4.64
0.100	0.100	9.28

Determine the rate law and the value of the rate constant.

3.   Derive the rate law for the reaction

$$BH^+ + A \longrightarrow AB + H^+$$

270

(a) If the mechanism is:

$$BH^+ \underset{}{\overset{K}{\rightleftharpoons}} B + H^+ \qquad \text{(fast)}$$

$$B + A \xrightarrow{\ k\ } AB \qquad \text{(slow)}.$$

(b) if the mechanism is:

$$BH^+ \xrightarrow{\ k\ } B + H^+ \qquad \text{(slow)}$$

$$B + A \longrightarrow AB \qquad \text{(fast)}$$

4. Differentiate between the <u>rate</u> of a chemical reaction and the <u>rate constant</u> for the reaction.

5. The half-lives for the reaction

$$A \longrightarrow P$$

at different concentrations are found to be as follows:

$[A]_i$	0.0015	0.0045	0.0090	0.0180
$t_{\frac{1}{2}}$ (sec)	7.6	2.5	1.3	0.63

Calculate (a) the order of the reaction and (b) the rate constant.

6. What is the fundamental difference between the half-time for a first order and second order reaction?

7. A compound decomposes as follows:

time (sec)	0	5	10	15	20	25	30
[A]	0.090	0.069	0.054	0.042	0.032	0.025	0.019

Determine the order and the rate constant.

8. A compound decomposes as follows:

time (sec)	0	20	40	60	80	100	160	200
[A]	0.286	0.274	0.261	0.250	0.240	0.231	0.206	0.192

Determine the order and the rate constant.

9. For the reaction

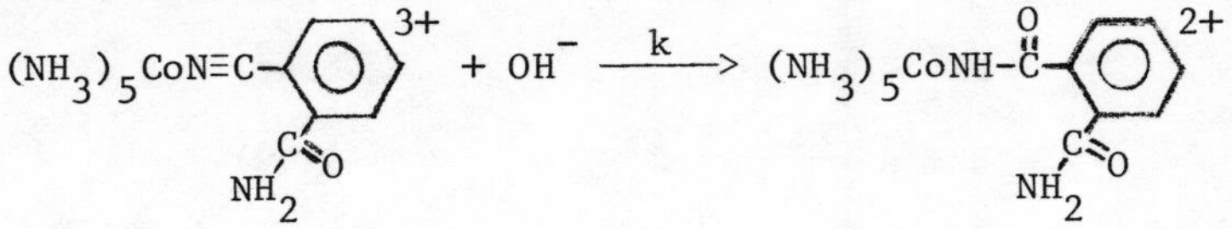

$$(NH_3)_5CoN\equiv C \text{—} \underset{\underset{NH_2}{\overset{\overset{C}{\shortparallel}}{}}}{\bigcirc}{}^{3+} + OH^- \xrightarrow{k} (NH_3)_5CoNH \text{—} \underset{\underset{NH_2}{\overset{\overset{C}{\shortparallel}}{}}}{\overset{\overset{O}{\shortparallel}}{C}} \bigcirc{}^{2+}$$

The following data were determined:

$T(°C)$	$k, M^{-1}s^{-1}$
25°	5.8×10^6
33°	1.6×10^7
38°	2.7×10^7
43°	4.7×10^7

Calculate E_a, the activation energy, for this reaction.

272

19

Selected Descriptive Chemistry of the Transition Metals

A. Points of Importance

1. New Terms and Ideas

ligand: a Lewis base that binds (coordinates) to a metal. eg. OH_2, NH_3, Cl^-, CN^-.

chelate: a ligand that coordinates to a metal through more than one donor atom. eg. $H_2N-CH_2-CH_2-NH_2$ (ethylenediamine), $^-O_2C-CO_2^-$ (oxalate), $H_3C-\underset{\underset{O}{\parallel}}{C}-CH=\underset{\underset{O_-}{}}{C}-CH_3$ (acetylacetonate).

monodentate ligand: ligand that can coordinate through one atom. eg. Cl^-, OH_2.

bidentate ligand: ligand that can coordinate through two atoms at once. eg. ethylenediamine, oxalate. Polydentate ligands can coordinate through 3,4,5,6.... atoms.

When a ligand combines with a metal ion a coordination complex (or complex ion) is formed. Some examples are shown on p. 274.

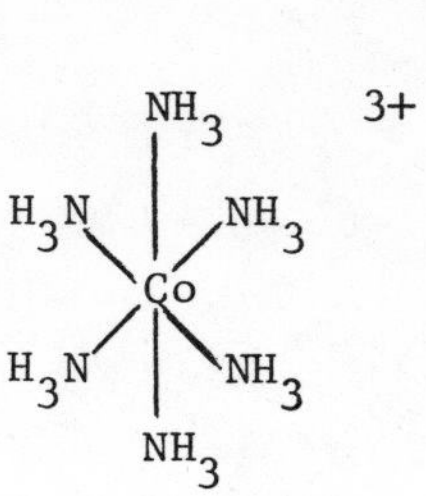

(a)   octahedral 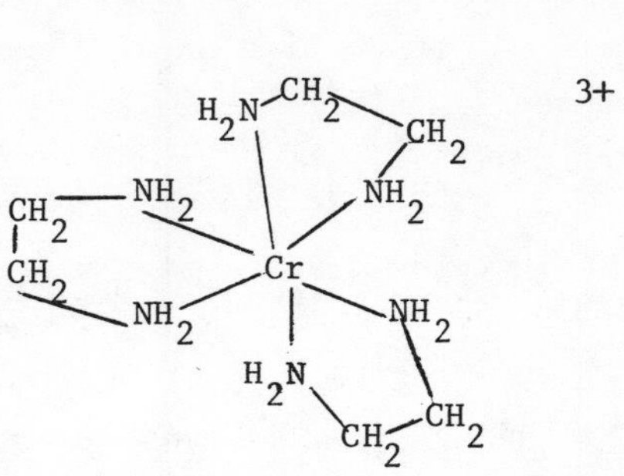   (b)   octahedral

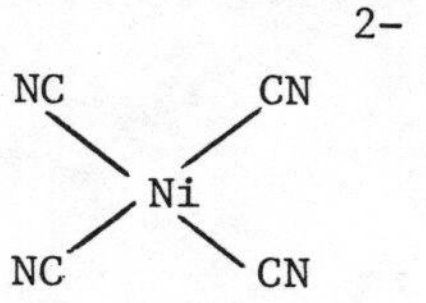

(c)   (d)   (e)

square planar   tetrahedral   square planar

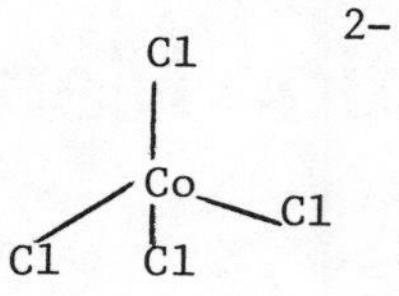

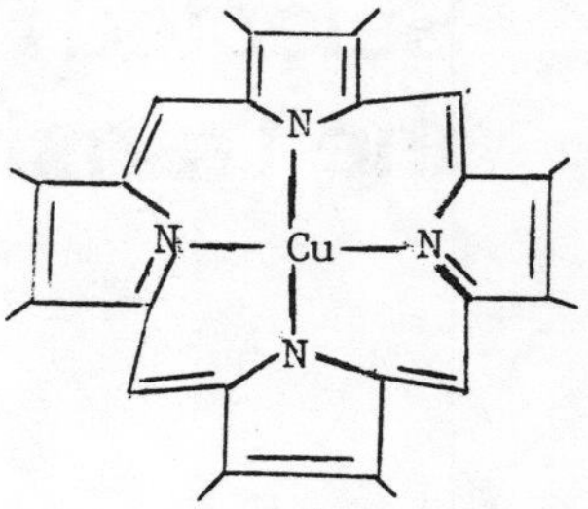

The coordination number refers to the total number of bonds formed
to the metal in the complex ion. In (a) above the coordination number
is 6 whereas in (c) and (d), it is 4. Notice that in (a) the overall
charge on the complex ion is +3. Since NH_3 is a neutral ligand this
means that the metal is in the +3 state. In (d) above the overall
charge on the complex ion is -2. Sinc Cl^- is a negatively charged
ligand the metal must be in the +2 state.

i.e. $4 \times Cl^- = -4$

                $1 \times Co^{2+} = \underline{+2}$

        overall charge = -2

274

Those ligands that are directly bonded to the metal are said to be in the inner or first coordination sphere. The species that balance this overall charge (maintain electroneutrality) are called the counterions. For example, in the solid state (a) can be isolated as the chloride or perchlorate salts:

$$\left[\begin{array}{c} \\ H_3N \underset{H_3N}{\overset{NH_3}{\underset{\displaystyle Co}{\diagup\!\!\!|\!\!\!\diagdown}}} \overset{NH_3}{\underset{NH_3}{\diagup\!\!\!|\!\!\!\diagdown}} \\ NH_3 \end{array} \right] Cl_3 \quad \text{and} \quad \left[\begin{array}{c} \\ H_3N \underset{H_3N}{\overset{NH_3}{\underset{\displaystyle Co}{\diagup\!\!\!|\!\!\!\diagdown}}} \overset{NH_3}{\underset{NH_3}{\diagup\!\!\!|\!\!\!\diagdown}} \\ NH_3 \end{array} \right] (ClO_4)_3$$

or

$[Co(NH_3)_6]Cl_3$ or

$[Co(NH_3)_6](ClO_4)_3$

In the above the NH_3 ligands are in the first coordination sphere whereas $3Cl^-$ and $3ClO_4^-$ are the counterions. The counterions are written outside the square brackets indicating that they are outside the first coordination sphere.

When the groups in the first coordination sphere are different, there is often more than one way to arrange these ligands about the metal. Complexes that have the same ligands arranged differently are called isomers. An example is given on the next page.

$[\text{Co structure with NH}_3, \text{H}_3\text{N}, \text{Cl}]$ Cl   and $[\text{Co structure with Cl, H}_3\text{N}, \text{NH}_3]$ Cl

cis: same groups next
to each other.

trans: same groups
opposite each
other.

2. Ligand Field Diagrams

The most important thing to remember in using ligand field diagrams as outlined in the text is the number of d electrons associated with a particular metal ion. You should know the following:

Metal	Notation	Ion	Notation
Ti	$3d^2 4s^2$	Ti^{3+}	$3d^1$
V	$3d^3 4s^2$	V^{2+}	$3d^3$
		V^{3+}	$3d^2$
Cr	$3d^5 4s^1$	Cr^{3+}	$3d^3$
Mn	$3d^5 4s^2$	Mn^{2+}	$3d^5$
Fe	$3d^6 4s^2$	Fe^{2+}	$3d^6$
		Fe^{3+}	$3d^5$
Co	$3d^7 4s^2$	Co^{2+}	$3d^7$
		Co^{3+}	$3d^6$
Ni	$3d^8 4s^2$	Ni^{2+}	$3d^8$
Cu	$3d^{10} 4s^1$	Cu^{2+}	$3d^9$
Pd	$4d^{10}$	Pd^{2+}	$4d^8$
Pt	$5d^9 6s^1$	Pt^{2+}	$5d^8$
		Pt^{4+}	$5d^6$

The ligand field diagrams for octahedral, tetragonal, square planar, and tetrahedral geometries are given below:

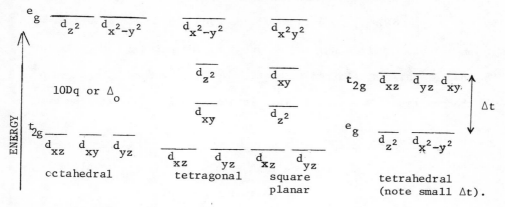

octahedral   tetragonal   square planar   tetrahedral (note small Δt).

(<u>Note</u>: Δtetrahedral is always much smaller than Δoctahedral). These orbitals contain the d electrons of the metal ion. The energy difference between the t_{2g} and e_g sets is called Δ or **10Dq**. The value of Δ is dependent on how strong the ligand binds to the metal. For example, consider the complexes $Co(OH_2)_6^{2+}$ and $Co(CN)_6^{4-}$. The CN^- ligand is a <u>strong field ligand</u> and binds very strongly to Co^{2+} and causes a large value of Δ. On the other hand, OH_2 is a <u>weak field ligand</u> and causes a small value of Δ. These facts are shown below:

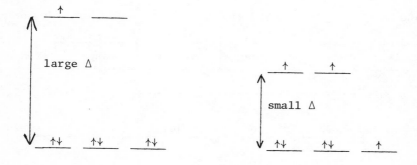

For $Co(CN)_6^{4-}$, Δ is large, and electron pairing results in the t_{2g} set before the upper e_g set is populated. This is referred to as <u>low spin</u> Co^{2+}. For $Co(OH_2)_6^{2+}$, Δ is small and each orbital is populated with one electron before pairing occurs i.e. the pairing energy is larger than Δ. This is referred to as <u>high spin</u> Co^{2+}. Each ligand has a characteristic value of Δ. This information is summarized below:

<u>weak field ligand</u>

F^-, OH_2 $\longrightarrow$ high spin $\longrightarrow$ small Δ

maximum number of
<u>unpaired</u> electrons

<u>strong field ligand</u>

CN^-, $H_2NCH_2CH_2NH_2$ $\longrightarrow$ low spin $\longrightarrow$ large Δ

maximum number of
<u>paired</u> electrons

When a complex absorbs light, an electron moves from the $t2g$ to the e_g orbitals and the color of the light <u>absorbed</u> depends on $\underline{\Delta}$. The color of the complex is the net of the colors <u>not absorbed</u>. This is illustrated below:

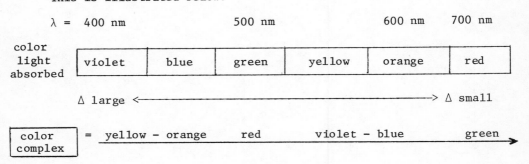

$\lambda =$ 400 nm		500 nm		600 nm	700 nm

color light absorbed

violet	blue	green	yellow	orange	red

Δ large $\longleftarrow$ $\longrightarrow$ Δ small

color complex = yellow – orange   red   violet – blue   green

B.   Types of Problems

1.   Inner and outer coordination sphere and the oxidation state of the
      metal.

2.   Prediction of isomerism given the geometry.

3.   Ligand field diagrams and their use.

Examples

1.   Question 7(c), (d), (e), and (f) at the end of the chapter.

      Calculate the oxidation state of the transition metal in the following
      complexes and indicate the charge on the species in the brackets.

      (c) $[Fe(CO)_3(PR_3)_2]$

      (d) $K_4[Fe(CN)_6]$

      (e) $K_3[Fe(CN)_6]$

      (f) $[Pt(NH_3)_2Cl_2]$.

(c) CO has zero charge as does PR_3

    Since the overall charge on the complex is zero the Fe is in the

    (0) state.

(d) Four potassium ions indicate an overall charge of 4− on $[Fe(CN)_6]$ i.e.
    $[Fe(CN)_6]^{4-}$. Within this unit we have 6 x −1(CN^-) = −6

$$\text{overall charge} = \underline{-4}$$
$$\therefore \text{ Fe is } \quad +2.$$

(e) Three potassium ions indicate an overall charge of −3 on $[Fe(CN)_6]$ i.e.
    $[Fe(CN)_6]^{3-}$. Within this unit we have 6 x −1 (CN^-) = −6

$$\text{overall charge} = \underline{-3}$$
$$\therefore \text{ Fe is } \quad +3.$$

279

(f) Within the unit $[Pt(NH_3)_2Cl_2]$ we have

$$2 \times 0 \ (NH_3) \ = \ 0$$

$$2 \times -1 \ (Cl^-) \ = \ -2$$

$$\text{overall charge} \ = \ \underline{\ 0 \ }$$

$$\therefore \ Pt \ \text{has} \qquad +2$$

It is not necessary to write everything out as above: a simple

procedure is as follows:

$$[Pt^{\textcircled{+2}}(NH_3)_2^0 Cl_2^{-2}]^0$$

Write the total charges above the ions within the square brackets and

make sure they add up to balance the overall charge. For problem (e)

above we have

$$K_3^{3+}[Fe^{\textcircled{+3}}(CN)_6^{-6}] \ \text{or} \ K_3^{3+} \ {}^{-3}[Fe(CN)_6].$$

2. Question 5. at the end of the chapter.

Predict all the possible isomers for

(a) $Co(NH_3)_5Cl,SCN,Br$ (six coordinate cobalt(III), Cl, Br, and SCN
     have a -1 oxidation state).

(b) $Ni(NH_3)_4Cl_2$ (four and six coordinate, with NH_3 always in the
     first coordination sphere).

(c) $Co(NH_3)_3Cl_3$ (six coordinate).

(d) $Fe(CO)_3[P(CH_3)_3]_2$ (trigonal bipyramid).

(e) $PtCl_3Br^{2-}$ (square planar).

(a) For NH_3 in the first coordination sphere there are no isomers, only

three different complexes:

$$[Co(NH_3)_5Cl](SCN)(Br)$$

$$[Co(NH_3)_5SCN](Cl)(Br)$$

$$[Co(NH_3)_5Br](SCN)(Cl).$$

(b) For $Ni(NH_3)_4Cl_2$ four coordinate we have only one compound and no isomers.

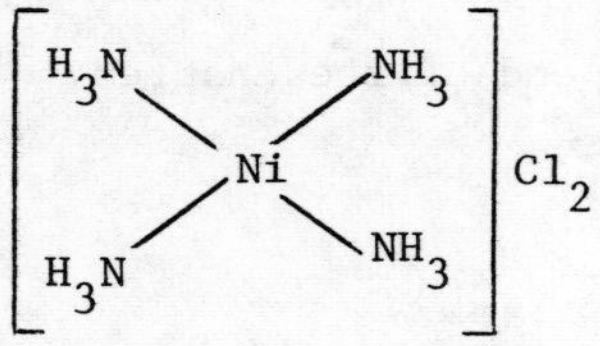

$$\left[\begin{array}{c} H_3N \qquad\qquad NH_3 \\ Ni \\ H_3N \qquad\qquad NH_3 \end{array}\right] Cl_2$$

For six coordinate cis and trans isomers are possible:

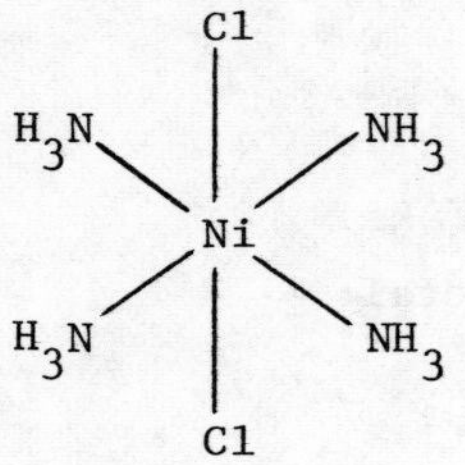

Cl — Ni with H₃N, H₃N, NH₃, NH₃ equatorial and Cl axial	NH₃ — Ni with H₃N, H₃N, Cl, Cl equatorial and NH₃ axial
trans	cis

(c)

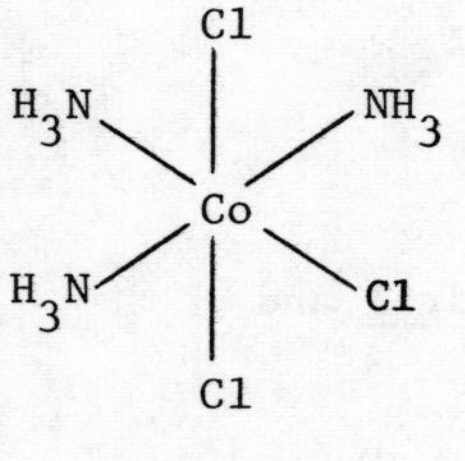

Cl(top)—Co—Cl(bottom); H₃N, H₃N, NH₃, NH₃	and	Cl(top)—Co—NH₃(bottom); H₃N, H₃N, Cl, Cl

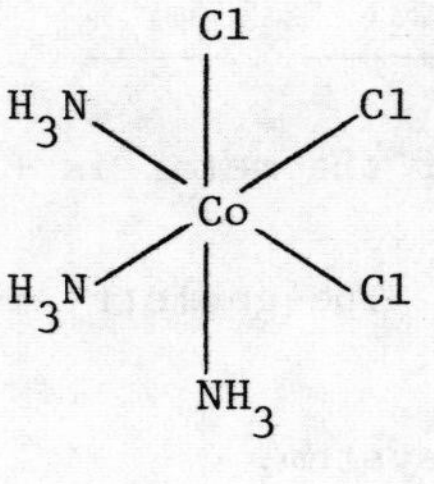

(d)

$$(CH_3)_3P,\ (CH_3)_3P,\ CO,\ CO,\ CO \ \text{about Fe}$$
$$OC,\ OC,\ P(CH_3)_3,\ P(CH_3)_3,\ CO\ \text{about Fe}$$
$$OC,\ OC,\ P(CH_3)_3,\ P(CH_3)_3,\ CO\ \text{about Fe}$$

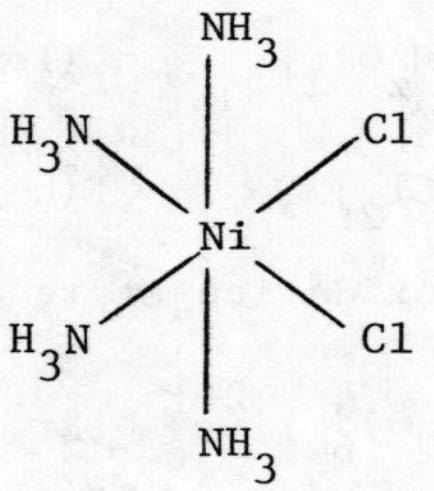

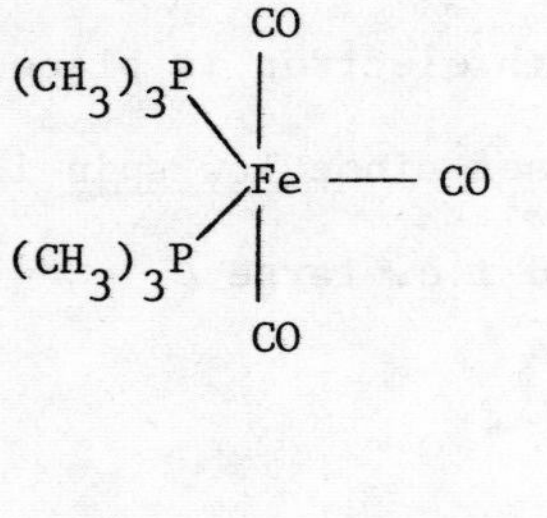

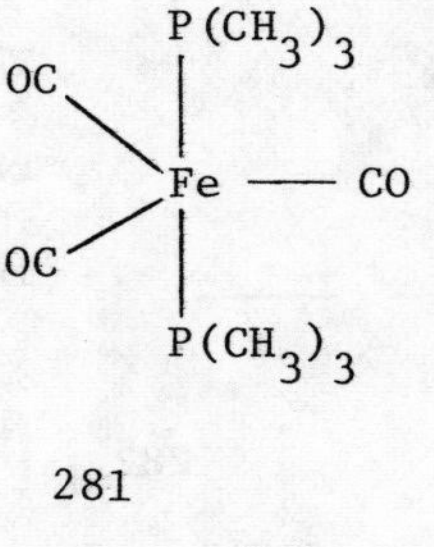

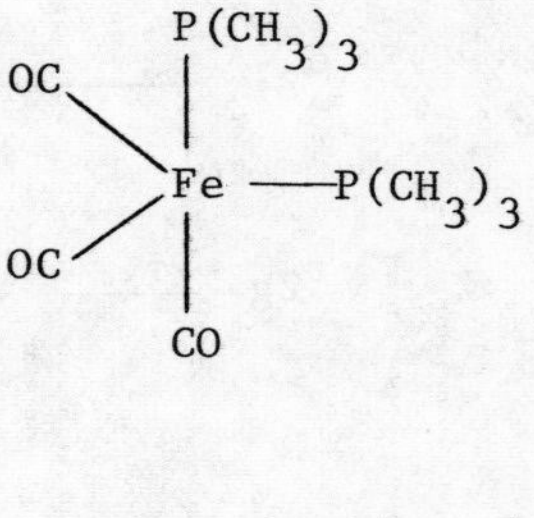

(e) There are no isomers of $PtCl_3Br^{2-}$ (square planar), only the complex

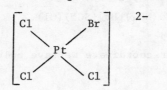

3.(i) Question 11(a), (c), (d), and (e) at the end of the chapter.

Draw the ligand field diagram for

(a) $[V(H_2O)_6]^{2+}$

(c) $[Cr(CN_6)]^{4-}$ (low spin)

(d) trans-$[Ni(CN)_4(H_2O)_2]^{2-}$ (low spin).

(e) trans-$[Ni(NH_3)_4Cl_2]$ (high spin).

(a) First determine the oxidation state of the metal:

$$[V^{\textcircled{2+}}(H_2O)_6^{0}]^{2+}$$

The complex is <u>octahedral</u> and V^{2+} is a d^3 system.

$$
\begin{array}{ll}
e_g & \underline{\quad} \quad \underline{\quad} \\
\\
t_{2g} & \underline{\uparrow} \quad \underline{\uparrow} \quad\quad \underline{\uparrow}
\end{array}
$$

(c) The oxidation state of the metal is +2:

$[Cr^{\textcircled{+2}}(CN)_6^{-6}]^{4-}$. The geometry is <u>octahedral</u> and

$\underline{Cr^{2+} \text{ is a } d^4}$ system.

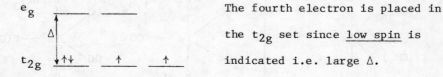

The fourth electron is placed in the t_{2g} set since <u>low spin</u> is indicated i.e. large Δ.

282

(d) The oxidation state of the metal is +2 (d^8):

trans $-$ [Ni$\overset{+2}{}$(CN)$_4^{-4}$(H$_2$O)$_2^0$]$^{2-}$. The <u>trans</u> geometry indicates

that the complex is <u>tetragonal</u>:

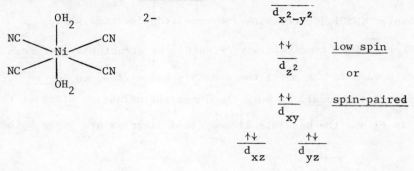

$$\overline{d_{x^2-y^2}}$$

$\underset{d_{z^2}}{\underline{\uparrow\downarrow}}$ <u>low spin</u>

      or

$\underset{d_{xy}}{\underline{\uparrow\downarrow}}$ <u>spin-paired</u>

$\underset{d_{xz}}{\underline{\uparrow\downarrow}}$ $\underset{d_{yz}}{\underline{\uparrow\downarrow}}$

If we assume that the Ni$-$OH$_2$ bonds are the same as the Ni$-$CN bonds

then we have octahedral geometry:

e_g $\underline{\quad\uparrow\quad}$ $\underline{\quad\uparrow\quad}$

t_{2g} $\underline{\quad\uparrow\downarrow\quad}$ $\underline{\quad\uparrow\downarrow\quad}$ $\underline{\quad\uparrow\downarrow\quad}$

(e) The oxidation state of the metal is +2 (d^8):

trans-[Ni$\overset{+2}{}$(NH$_3$)$_4^0$Cl$_2^{2-}$]. Again tetragonal geometry or

octahedral geometry is indicated:

$\underline{\quad\uparrow\quad}$ }
        }   <u>high spin</u>
$\underline{\quad\uparrow\quad}$ }

$\underline{\quad\uparrow\downarrow\quad}$

$\underline{\quad\uparrow\downarrow\quad}$ $\underline{\quad\uparrow\downarrow\quad}$

       tetragonal

$\underline{\quad\uparrow\quad}$ $\underline{\quad\uparrow\quad}$

$\underline{\quad\uparrow\downarrow\quad}$ $\underline{\quad\uparrow\downarrow\quad}$ $\underline{\quad\uparrow\downarrow\quad}$ <u>low spin</u>

       octahedral

<u>high spin</u> or

Note that for octahedral geometry both d^8 low spin and high spin give

the same ligand field diagrams. Tetragonal geometry on the other hand

283

has 2 unpaired electrons for a d^8 high spin complex and 0 unpaired electrons for a d^8 low spin system.

(ii) Question 12 at the end of the chapter.

The complex $[NiCl_4]^{2-}$ contains two unpaired electrons while $[Ni(CN)_4]^{2-}$ doesn't contain any. Predict the structures. For <u>four</u> ligands bonded to the metal the possible geometries are <u>square planar</u> or <u>tetrahedral</u>. In both complexes the oxidation state of the metal is +2 and the possible ligand field diagrams are given below:

```
  ↑    ___          ___                       ___

  ↑    ___        ↑↓ ___   ↑↓ ↑   ↑   ___    ↑↓ ___   ↑↓ ↑   ↑   ___

 ↑↓ ___          ↑↓ ___              ↑↓ ___

 ↑↓   ↑↓         ↑↓   ↑↓   ↑↓   ↑↓   ↑↓   ↑↓   ↑↓   ↑↓
 ___  ___        ___  ___  ___  ___  ___  ___  ___  ___
(high spin)     (low spin)                square        tetrahedral
square          square       tetrahedral  planar
planar          planar
```

$$[NiCl_4]^{2-} \qquad\qquad\qquad [Ni(CN)_4]^{2-}$$

If we assume the $[NiCl_4]^{2-}$ complex is high spin (Cl^- is a weak field ligand) then both square planar and tetrahedral geometries are possible. However most square planar complexes of Ni^{2+} are low spin and all electrons are paired. Thus $[NiCl_4]^{2-}$ is most likely <u>tetrahedral</u> (2 unpaired electrons). Since CN^- is a strong field ligand, $[Ni(CN)_4]^{2-}$ is low spin. Since the complex has no unpaired electrons the complex must be square planar.

(iii) Question 15. at the end of the chapter.

The complex $[TiL_6]^{3+}$ is violet; $[TiR_6]^{3+}$ is yellow. Which complex has the larger Δ? In this case, the complexes are octahedral and the oxidation sate of titanium is +3 (d^1):

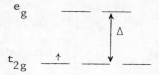

The color is caused by what is left over after the light is absorbed corresponding to excitation of the one electron from $t_{2g} \rightarrow e_g$ (Δ). Thus for $[TiL_6]^{3+}$ yellow-orange light is absorbed giving a violet complex. Note that this color corresponds to a small Δ. $[TiR_6]^{3+}$ is yellow thus the light absorbed was violet-blue. This color corresponds to a large Δ.

Therefore yellow $[TiR_6]^{3+}$ has a larger Δ than violet $[TiL_6]^{3+}$. The table at the end of section A. 2. summarizes these ideas.

(iv) Question 17. at the end of the chapter.

A complex of nickel(II) is found to contain four NH_3 molecules and two chloride ions. It contains two unpaired electrons. What structures are eliminated and which ones are possible?

For $4NH_3$ and $2Cl^-$ we have two possible coordination numbers: 4 and 6. These are $[Ni(NH_3)_4]Cl_2$ and $[Ni(NH_3)_4Cl_2]$. For the first formulation <u>square planar</u> and <u>tetrahedral</u> geometries are possible while <u>octahedral</u> geometry is required by the latter formulation. The ligand field diagrams for these possibilities are given below: (assume low spin for square planar).

↑↓	↑↓ ↑ ↑	↑ ↑
↑↓	↑↓ ↑↓	
↑↓ ↑↓		↑↓ ↑↓ ↑↓
square planar	tetrahedral	octahedral
(a)	(b)	(c)

Since this complex has 2 unpaired electrons we can eliminate (a).

Both (b) and (c) are possible.

C. Test Yourself

1. Calculate the oxidation state of the transition metal in the following complexes.

   (a) $NH_4[Cr(NH_3)_2(NCS)_4]$

   (b) $[Pt(NH_3)_4](PtCl_4)$

   (c) $K[Co(NH_3)_2(NO_2)_4]$

   (d) $[Ru(OH_2)_4(Cl)_2]Br$

   (e) $Na_4[NiCl_6]$

2. Silver nitrate was added to solutions of the following octahedral complexes and AgCl was precipitated immediately in the amounts indicated.

Complex Formula	Moles of AgCl/Mole of Complex
$CoCl_3(NH_3)_6$	3
$CoCl_3(NH_3)_5$	2
$CoCl_3(NH_3)_4$ (purple)	1
$CoCl_3(NH_3)_4$ (green)	1

   (a) Draw the structures expected for each of these complexes.

   (b) Explain the fact that $CoCl_3(NH_3)_4$ can be purple or green but both forms give 1 mole of AgCl per mole of complex.

3. The complex $Ni(CN)_4^{2-}$ is yellow in color while $Ni(OH_2)_6^{2+}$ is green. Which complex has the greatest average separation of the d-orbitals?

4. For the complexes $TiCl_6^{3-}$ and $Ti(CN)_6^{3-}$ which has the greater value of Δ? What color might you expect the complexes to be to the eye?

5. For the complex $Co(NH_3)_2$ (pyridine)$_2Cl_2$ predict the structures for

287

octahedral, square planar, and tetrahedral geometries. How could you differentiate between these possibilities?

6. Draw ligand field diagrams for the following:

    (a) $[Fe(OH_2)_6]^{2+}$ and $[Fe(CN)_6]^{4-}$.

    (b) $[Cr(OH_2)_6]^{3+}$ and $[Cr(OH_2)_6]^{2+}$.

    (c) $[Co(NO_2)_6]^{3-}$ and $[Mn(OH_2)_6]^{3+}$.

7. How many unpaired electrons would you predict for the following complexes.

    (a) $Cu(CN)_4^{3-}$.

    (b) $NiCl_4^{4-}$.

    (c) $CoCl_4^{2-}$.

8. Draw all possible geometric isomers of

    (a) $[Co(en)_2(NH_3)Cl]^{2+}$

    (b) $[Co(en)(NH_3)_2Cl_2]^{+}$.

20

Selected Descriptive Chemistry of the Main Group Elements

A.   <u>Points of Importance</u>

This chapter covers a great deal of descriptive chemistry of groups I, II, III, IV, V, VI, and VIIA. The elements are broken up on the periodic table below according to metals, nometals (shaded area), and metalloids (hatched area). It is important to know the

Li	Be											B	C	N	O	F	
Na	Mg											Al	Si	P	S	Cl	
K	Ca											Ga	Ge	As	Se	Br	
Rb	Sr											In	Sn	Sb	Te	I	
Cs	Ba											Tl	Pb	Bi	Po	At	
Fr	Ra																

location of these classes of main group elements on the periodic table. The three classifications become especially important when considering group trends within groups III, IV, V and VIA.

1. Groups IA and IIA.

In addition to the properties summarized in chapter 10 these two groups are distinguished for their organometallic chemistry and their participation in biochemical processes. Organometallic compounds are those that contain a metal-carbon bond. Examples are sodium propide, $Na^+CH(CH_3)_2^-$, and Grignard reagents, RMgX, where R can be an alkyl group and X a halide. The cations Na^+, Li^+, Mg^{2+}, and Ca^{2+} are essential in a number of control and trigger mechanisms in the body. For example, Mg^{2+} is required to activate a great number of enzymes essential in biosynthesis and control of nerve impulses.

2. Group IIIA.

In this group the boron hydrides form a unique group of compounds. The simplest boron hydride is diborane, B_2H_6, which has the following structure

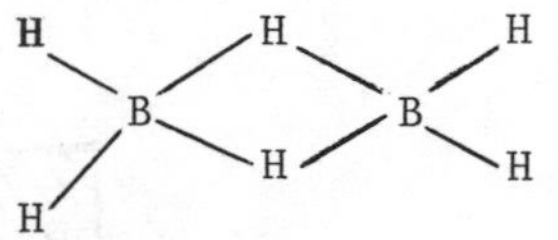

As pointed out in the text diborane is an electron deficient compound since there are more bonding orbitals available than electrons to form electron pair bonds using these orbitals. The bonding is explained in terms of two 3-center bonds for the bridging hydrogens. Diborane is

290

extremely reactive: two representative reactions are given below.

$$B_2H_6 + 6H_2O \longrightarrow 2B(OH)_3 + 6H_2$$

$$B_2H_6 + 2(CH_3)_3N \longrightarrow 2(CH_3)_3NBH_3$$

3. **Group IVA**.

C   nonmetallic, forms bonds to itself (catenation), important
    compounds are CO, CO_2 and organic molecules.

Si metalloid, silicon dioxide ,SiO_2, most important unit.

Ge metalloid, important compounds are GeO, GeO_2, GeH_4. Both 2+
    and 4+ species, mostly 4+.

Sn metal, both 2+ and 4+ species known: SnO_2, SnH_4, SnO, SnS_2.

Pb metal, only a few 4+ compounds important, PbO_2, $PbCl_4$; mostly 2+
    compounds: PbO, $PbCl_2$, $Pb(NO_3)_2$.

4. **Group VA**.

These elements offer a wide variety of properties from the non-
metallic N to metallic Bi through the metalloids As and Sb. The most
important member of this group is N which exhibits a great diversity
in its compounds and reactions:

$$2NO(g) + O_2(g) \rightarrow \underline{2NO_2}(g) \quad \text{(air pollution)}$$

$$NH_4^+(aq) + NO_2^-(aq) \rightarrow \underline{N_2}(g) + 2H_2O \quad \text{(laboratory preparation}$$
$$\text{of pure } N_2)$$

$$\underline{NCl_3}(\ell) + 3H_2O \rightarrow NH_3(g) + 3HOCl(aq) \quad \text{(hydrolysis of halides)}$$

$$3NO_2(g) + H_2O \rightarrow \underline{2H^+}(aq) + \underline{2NO_3^-}(aq) + NO(g) \quad \text{(preparation of}$$
$$\text{nitric acid)}$$

5. **Group VIA**.

The chemistry of this group is dominated by the nonmetallic
elements O, S, and Se.

291

<u>Oxygen</u>: exists in two gaseous, allotropic forms, O_2 and O_3.

<u>Sulfur</u>: stable form of sulfur is a ring structure, S_8.

Burning (roasting) of S-containing ores or elemental S produces the pollutant SO_2.

$$4FeS_2(s) + 11O_2(g) \rightarrow 2Fe_2O_3(s) + 8SO_2(g)$$

$$2PbS(s) + 3O_2(g) \rightarrow 2PbO(s) + 2SO_2(g)$$

The SO_2 can be prevented from reaching the atmosphere by <u>converting it to sulfuric acid</u> or by the reaction

$$CaO(s) + SO_2(g) \rightarrow CaSO_3(s).$$

<u>Selenium</u>: stable form of Se consists of chains of atoms. Element is used in the construction of photoelectric cells.

Again metallic character increases as one moves down the group.

6. <u>Group VIIA</u> – the Halogens (all nonmetals).

F: pale yellow F_2 <u>gas</u> at room temperature; F_2 is the strongest oxidizing agent in water in this group because of the small size of the F^- ion which leads to a larger solvation energy than the other members of the group; $r(F^-) = 1.36$ Å. Acid, HF, prepared by $CaF_2(s) + H_2SO_4(\ell) \rightarrow CaSO_4(s) + 2HF(g)$.

Cl: yellow-green Cl_2 <u>gas</u> at room temperature; $r(Cl^-) = 1.81$Å; bond energy $Cl_2 = 58$ kcal mole^{-1}. Acid, HCl, prepared by $NaCl(s) + H_2SO_4(\ell) \rightarrow NaHSO_4(s) + HCl(g)$.

Br: red-brown Br_2 <u>liquid</u> at room temperature; $r(Br^-) = 1.95$ Å; bond energy $Br_2 = 46$ kcal mole^{-1}. Acid, HBr, prepared by $NaBr(s) + H_3PO_4(\ell) \rightarrow HBr(g) + NaH_2PO_4(s)$.

292

I: violet-black I_2 __solid__ at room temperature; $r(I^-) = 2.16$ Å;

bond energy $I_2 = 36$ kcal mole^{-1}. Acid, HI, prepared by

$$NaI(s) + H_3PO_4 (\ell) \rightarrow HI(g) + NaH_2PO_4(s).$$

The phosphorus trihalides all react with water to give the

appropriate acid. For example,

$$PCl_3 + 3H_2O \rightarrow 3HCl(g) + H_3PO_3(aq).$$

What type of interaction occurs between PCl_3 and H_2O?

Another important series of compounds formed by the halogens

are the oxyacids. The oxyacids for chlorine are listed below

along with their oxidation states:

	oxidation state Cl
HOCl	+1
$HClO_2$	+3
$HClO_3$	+5
$HClO_4$	+7

The standard electrode potentials for Cl_2 in aqueous acid are given

below:

$$ClO_4^- \xrightarrow{1.19v} ClO_3^- \xrightarrow{1.21v} HClO_2 \xrightarrow{1.64v} HOCl \xrightarrow{1.63v} Cl_2 \xrightarrow{1.36v} Cl^-$$

$$\underset{1.47v}{\underbrace{\hspace{7cm}}}$$

The numbers shown on the arrow are for the reduction of the species

on the left to that on the right. Since all the values are positive,

the oxyhalides are all good oxidizing agents. For example hypo-

chlorite, OCl^-, is used as a bleach. Whenever a species in the above diagram has a more positive potential on the right (reduction to a lower oxidation state) than on the left (oxidation to a higher oxidation state) that substance is unstable toward <u>disproportiona-tion</u>. For example, for the series

$$ClO_3^- \xrightarrow{\text{1.21v}} HClO_2 \xrightarrow{\text{1.64v}} HOCl$$

1.64v (to the right of $HClO_2$) is more positive than + 1.21v (to the left of $HClO_2$) and $HClO_2$ will disproportionate:

$$HClO_2 + H_2O \longrightarrow ClO_3^- + 3H^+ + 2e^- \qquad\qquad E° = -1.21v$$

$$2H^+ + HClO_2 + 2e^- \longrightarrow HOCl + H_2O \qquad\qquad E° = +1.64v$$

$$2HClO_2 \longrightarrow ClO_3^- + HOCl + H^+ \qquad\qquad E° = +0.23v$$

Note that for the reaction written from right to left ($HClO_2 \rightarrow ClO_3^-$) the sign of E° was changed. Remember this type of reaction is called <u>disproportionation</u>.

Redox diagrams are extremely important when predicting if reactions will occur and whether species are stable in acid or base. Diagrams of the type given above can also be prepared for O, N, P, As, Br, and I. For example, a simplified redox diagram for O in acidic solution is given below:

$$O_2 \xrightarrow{\text{0.67v}} H_2O_2 \xrightarrow{\text{1.77v}} H_2O$$
$$\big|\underline{\qquad\qquad \text{1.23v} \qquad\qquad}\big|$$

This diagram indicates that H_2O_2 disproportionates to O_2 and H_2O.

B.   Types of Problems

1.   Problems involving the use of redox potentials within a group or
      within a series of related compounds of the same element.

2.   Problems involving descriptive chemistry.

Examples

1.(i) Question 4. at the end of the chapter.

      Write a balanced equation for the reaction of Br^- with Cl_2.

          From Table 20-7 we see that the $E°$'s for $\frac{1}{2} X_2(g) + e^- \rightarrow X^-(aq)$

          decrease as we go from F_2 to I_2. This means that a halide at

          the top of the group will oxidize the anion of any halogen

          below it. Therefore Cl_2 will oxidize Br^-:

              $$Cl_2(g) + 2Br^-(aq) \rightarrow 2Cl^-(aq) + Br_2(\ell).$$

(ii) For the following reduction potential diagram, predict which

      compounds disproportionate.

          $$SO_4^{2-} \xrightarrow{0.17v} \boxed{SO_2} \xrightarrow{0.40v} H_2S_2O_3 \xrightarrow{0.50v} S_8 \xrightarrow{0.14v} H_2S$$

      Using the rule that if species have larger reduction potentials on the

      right than on the left (i.e. going from right to left diagramatically)

      we predict that SO_2 will disproportionate to form SO_4^{2-} and $H_2S_2O_3$.

      Also $H_2S_2O_3$ will disproportionate to form S_8 and SO_2.

(iii) Question 21. at the end of the chapter.

      There are no readily available chemical reactions for converting F^-

      to F_2. Explain why. F_2 is one of the strongest oxidizing agents known:

              $$e^- + \frac{1}{2} F_2(g) \longrightarrow F^-(aq), \quad E° = +2.87v$$

      This means that it is very unfavourable for F^- to go to F_2.

      See section 20-7 in text.

(iv) Question 23. at the end of the chapter.

HBr cannot be prepared by the action of H_2SO_4 with NaBr although HF can be prepared by the reaction of H_2SO_4 with NaF. Explain.

Since F^- is very difficult to oxidize $(e^- + \frac{1}{2}F_2(g) \rightarrow F^-(aq), \; E^\circ = +2.87v)$ sulfuric acid can act in a simple displacement reaction to produce HF:

$$NaF(s) + H_2SO_4(\ell) \rightarrow NaHSO_4(s) + HF(g)$$

On the other hand, Br^- is not as difficult to oxidize $(e^- + \frac{1}{2}Br_2(\ell) \rightarrow Br^-(aq), \; E^\circ = +1.06v)$ and sulfuric acid oxidizes Br^- to Br_2:

$$2NaBr(s) + 2H_2SO_4(\ell) \rightarrow Br_2(g) + SO_2(g) + Na_2SO_4(s) + 2H_2O(g)$$

Phosphoric acid can be used instead of sulfuric:

$$NaBr(s) + H_3PO_4(\ell) \rightarrow HBr(g) + NaH_2PO_4(s).$$

2. Predict the products of the following reactions:

(a) $PCl_3(\ell) + Cl_2(g) \longrightarrow$

(b) $PCl_3(\ell) + H_2O(\ell) \longrightarrow$

(c) $GaCl_3(s) + H_2O(\ell) \longrightarrow$

(d) $NO_2(g) + H_2O(\ell) \longrightarrow$

(e) $SO_3(g) + H_2SO_4(\ell) \longrightarrow$

(a) $PCl_3(\ell) + Cl_2(g) \longrightarrow PCl_5(s)$

(b) $PCl_3(\ell) + 3H_2O(\ell) \longrightarrow H_3PO_3(aq) + 3H^+(aq) + 3Cl^-(aq)$

(c) $GaCl_3(s) + 3H_2O(\ell) \longrightarrow Ga(OH)_3(s) + 3H^+(aq) + 3Cl^-(aq)$

(d) $3NO_2(g) + H_2O(\ell) \longrightarrow 2HNO_3(aq) + NO(g).$

(e) $SO_3(g) + H_2SO_4(\ell) \longrightarrow H_2S_2O_7(\ell).$

C. Test Yourself

1. Predict the products of the following reactions. If no reaction occurs write NR.

   (a) $Br_2(\ell) + 2NaI(s) \longrightarrow$

   (b) $Br_2(\ell) + 2NaCl(s) \longrightarrow$

   (c) $NaCl(s) + H_3PO_4(\ell) \longrightarrow$

   (d) $ZnS(s) + O_2(g) \longrightarrow$

   (e) $P_4O_{10}(g) + C(s) \longrightarrow$

   (f) $AsCl_3(\ell) + H_2O \longrightarrow$

   (g) $NaNO_3(s) + H_2SO_4(\ell) \longrightarrow$

   (h) $Cl_2(g) + OH^-(aq) \longrightarrow$

   (i) $As_2O_3(s) + 3C(s) \longrightarrow$

2. For the following reduction potential diagram, predict which compounds disproportionate.

$$NO_3^- \xrightarrow{0.94v} HNO_2 \xrightarrow{1.00v} NO \xrightarrow{1.59v} N_2O \xrightarrow{1.77v} N_2 \xrightarrow{0.27} NH_3$$
$$\vert \underline{\hspace{3cm} 1.11v \hspace{3cm}} \vert$$

3. Write the three equations responsible for the production of nitric acid by the Ostwald process. The last reaction involves the disproportionation of $NO_2(g)$. Explain what this means in light of the products.

4. (a) What are the products of the thermal decomposition of $NH_4NO_3(s)$?

   (b) Write the equation for the reaction of $P_4O_6(s)$ with H_2O.

   (c) What are the products of the reduction of $Sb_2O_3(s)$ by $H_2(g)$?

   (d) What are the products of the thermal decomposition of $NH_4NO_2(s)$?

5. (a) Can $Cl_2(g)$ be used to prepare $Br_2(\ell)$ from NaBr?

(b) Can $Cl_2(g)$ be used to prepare $F_2(g)$ from NaF?

(c) What are the products of the disproportionation of $S_2O_3^{2-}$ in acidic solution?

6. Given the following redox diagrams predict whether elemental phosphorus and arsenic are stable in acidic solution.

$$H_3PO_4 \xrightarrow{-0.28} H_3PO_3 \xrightarrow{-0.50} H_3PO_2 \xrightarrow{-0.51} P_4 \xrightarrow{-0.06} PH_3$$

$$H_3AsO_4 \xrightarrow{+0.56} H_3AsO_3 \xrightarrow{+0.25} As \xrightarrow{-0.60} AsH_3.$$

21

Chemistry of the Hydrocarbons

A. Points of Importance

1. New Terms

Rotamers: different arrangements of atoms that can be converted into

one another by rotation about single bonds. eg. Using

Newman projections for propane, C_3H_8, we have

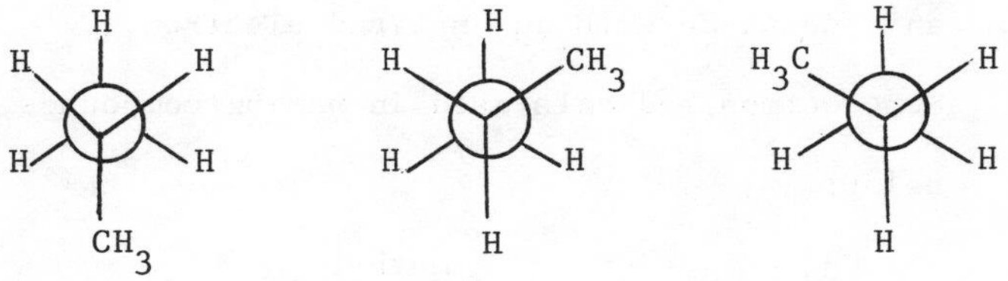

These representations are <u>indistinguishable conformations</u>

of the same molecule and are called <u>rotamers</u>.

Structural Isomers: compounds with the same molecular formula but

different structures. eg. for pentane, C_5H_{12}, there are 3 structural isomers:

```
      H   CH₃  H    H                    H                  H   H   H   H   H
      |   |    |    |                    |                  |   |   |   |   |
  H - C - C -  C  - C - H            H-C-H             H  - C - C - C - C - C - H
      |   |    |    |               H₃C-C-CH₃              |   |   |   |   |
      H   H    H    H                   H-C-H               H   H   H   H   H
                                          |
                                          H

        (a)                            (b)                          (c)
```

These compounds are all different but have the same molecular formula
C_5H_{12}.

```
┌─────────────────────────────────────────────────────────────────┐
│                                                                   │
│  To distinguish between structural isomers and rotamers, as well as │
│                                                                   │
│  visualizing stereochemistry molecular models are indispensable and │
│                                                                   │
│  should be used.                                                  │
│                                                                   │
└─────────────────────────────────────────────────────────────────┘
```

For example, verify the different structural isomers given above by

building molecular models of each. For isomer (c) rotate each C - C

bond: notice you still have the same compound. In order to make (a)

or (b) from (c) it is necessary to break a bond and make a new bond.

Try it!

Radical: any substance with an unpaired electron.

          Some common radicals used in naming compounds are given

          below:

          CH_3 methyl

          CH_3CH_2 ethyl

          $CH_3CH_2CH_2$ propyl

$$H_3C$$
$$H_3C$$
$$CH$$ isopropyl

These are in general called alkyl groups.

Alkanes: C_nH_{2n+2}, saturated hydrocarbons. Saturated simply means
           that each carbon has four single bonds either to hydrogens
           or carbons.

Alkenes: C_nH_{2n}, unsaturated hydrocarbons containing at least one
           double bond. Unsaturated refers to the fact that there is
           a carbon-carbon double bond i.e. the bonds are not all
           single bonds.

Alkynes: C_nH_n, unsaturated hydrocarbons containing at least one
           triple bond.

Addition Reaction: two molecules combine to yield a single molecule
           of product.

Electrophilic Reagents: acidic reagents that are seeking a pair of
           electrons i.e. Lewis acids.

Nucleophilic Reagents: Lewis bases.

Markovnikov's Rule: ionic addition of an acid to the carbon-
           carbon double bond of an unsymmetrical alkene occurs
           by addition of the hydrogen of the acid to the carbon
           atom of the alkene that already holds the greater
           number of hydrogens.

Polymerization: the joining together of small molecular units to make
           a very large molecule.

301

<u>Carbonium Ion</u>: a group of atoms that contains a carbon atom with only six electrons. eg.

$$
\underset{\text{methyl cation}}{H\!:\!\overset{\displaystyle H}{\underset{\displaystyle H}{C}}\!\oplus \text{ or } H\!-\!\overset{\displaystyle H}{\underset{\displaystyle H}{C}}\!\oplus}
\qquad
\underset{\text{ethyl cation}}{CH_3\!-\!\overset{\displaystyle H}{\underset{\displaystyle H}{C}}\!\oplus}
$$

<u>Aromatic Compounds</u>: benzene, C_6H_6, and compounds that resemble benzene in chemical behaviour.

2. <u>Reactions of Alkanes</u>

(a) combustion

$$C_8H_{18}(\ell) + \frac{25}{2}\,O_2(g) \longrightarrow 8CO_2(g) + 9H_2O(\ell)$$

(b) cracking

$$\text{alkane} \xrightarrow{\;700^\circ\;} H_2(g) + \text{smaller alkanes} + \text{alkenes}$$

(c) halogenation (free radical mechanism).

$$CH_3CH_3 \xrightarrow[\text{light}]{Cl_2} CH_3CH_2Cl$$

3. <u>Reactions of Alkenes</u>

(a) ionic addition (Markovnikov addition).

$$CH_3CH{=}CH_2 \xrightarrow{\;HBr\;} CH_3\underset{\displaystyle Br}{C}HCH_3$$

(b) addition of halogens

$$CH_3CH{=}CH_2 \xrightarrow[\text{in } CCl_4]{Br_2} CH_3\underset{\displaystyle Br}{C}HCH_2Br$$

(c) free radical addition (non-Markovnikov addition)

$$CH_3CH=CH_2 \xrightarrow[\text{peroxide}]{\text{HBr}} CH_3CH_2CH_2Br$$

(free radical mechanism)

(d) addition of hydrogen

$$CH_3CH=CH_3 \xrightarrow[\substack{Ni \\ \text{catalyst}}]{H_2} CH_3CH_2CH_3$$

(e) hydroxylation (oxidation of double bonds).

$$CH_3CH=CH_2 \xrightarrow{KMnO_4} CH_3\underset{OH}{CH}\ \underset{OH}{CH_2}$$

(f) free radical polymerization

$$nCH_2=CCl_2 \xrightarrow{\text{peroxides}} \sim\sim\sim\sim CH_2-CCl_2-CH_2-CCl_2-CH_2-CCl_2\sim\sim\sim\sim$$

saran
(packaging film)

(g) addition of sulfuric acid

$$CH_3CH=CH_2 + H_2SO_4 \longrightarrow CH_3\underset{OSO_3H}{CH}-CH_3$$

4. Reactions of Alkynes

(a) combustion

$$CH{\equiv}CH(g) + \frac{5}{2}O_2(g) \longrightarrow 2CO_2(g) + H_2O(g)$$

(b) many addition reactions

$$-C{\equiv}C- \ +\ XY \longrightarrow \underset{X\ \ Y}{-C=C-} \xrightarrow{XY} \underset{X\ \ Y}{\overset{X\ \ Y}{-C-C-}}$$

where XY can be H_2, Cl_2, HCN etc.

(eg.) $H-C{\equiv}C-H + HCl \longrightarrow \underset{H\ \ Cl}{H-C=C-H}$ vinyl chloride

5. <u>Reactions of Aromatic Hydrocarbons</u> (Benzene)

Electrophilic aromatic substitution:

<u>examples:</u>

<u>halogenation:</u>

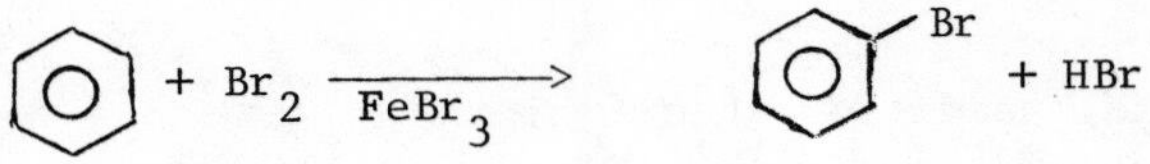

$\bigcirc$ + Br_2 $\xrightarrow{FeBr_3}$ $\bigcirc$—Br + HBr

<u>nitration:</u>

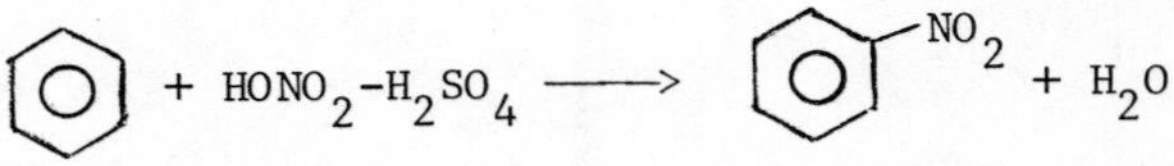

$\bigcirc$ + $HONO_2$ $-H_2SO_4$ $\longrightarrow$ $\bigcirc$—NO_2 + H_2O

<u>sulfonation:</u>

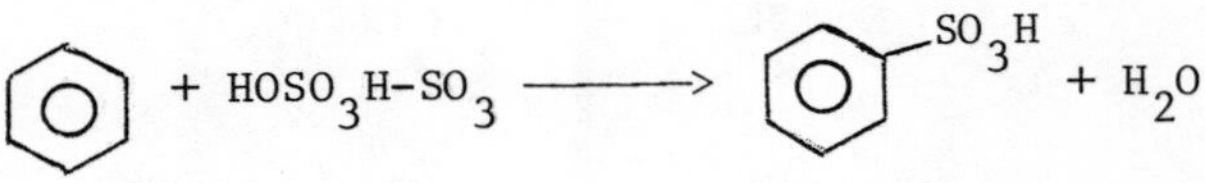

$\bigcirc$ + $HOSO_3H$-SO_3 $\longrightarrow$ $\bigcirc$—SO_3H + H_2O

<u>Friedel-Crafts alkylation</u> (AlX_3-CH_3X)

$\bigcirc$ + CH_3Cl $\xrightarrow{AlCl_3}$ $\bigcirc$—CH_3 + HCl

6. <u>Directive Influence of Substituents on Electrophilic Aromatic Substitution</u>

<u>electron releasing substituents</u> - favor <u>ortho</u>, <u>para</u> attack.

Alkyl groups, NH_2, OH, OCH_3.

<u>electron withdrawing substituents</u> - favor <u>meta</u> attack

NO_2, CN, CF_3, SO_3H, CO_2H.

<u>halogens</u> - electron withdrawing but <u>ortho</u>, <u>para</u> - directing.

These should be considered a special class.

B.   Types of Problems

1.   Structural isomers and nomenclature of <u>alkanes</u>.

2.   <u>Alkenes</u>, <u>cis</u> and <u>trans</u> isomers, and nomenclature.

3.   Structures and nomenclature of <u>aromatic</u> hydrocarbons.

4.   Reactions of alkanes.

5.   Reactions of alkenes.

6.   Reactions of benzene.

<u>Examples</u>

1.(i) Question 10. at the end of the chapter.

Draw the structural formula and give the IUPAC name for

(a) $(CH_3)_3CCH_2C(CH_3)_3$.

(b) $(C_2H_5)_2C(CH_3)CH_2CH_3$.

(c) $CH_3CH_2C(CH_3)_2CH_2CH_3$.

(a) (1) Draw the structure and pick out the longest carbon chain.

(2) Number the carbons in the longest chain.

$$H_3\underset{1}{C}-\underset{2}{\overset{\overset{\displaystyle CH_3}{|}}{\underset{\underset{\displaystyle CH_3}{|}}{C}}}-\underset{3}{CH_2}-\underset{4}{\overset{\overset{\displaystyle CH_3}{|}}{\underset{\underset{\displaystyle CH_3}{|}}{C}}}-\underset{5}{CH_3}$$

(3) Name the groups attached to carbons of the longest chain and
indicate their position by the number in the longest chain and
indicate the number of groups at that position by a prefix.

(4) The parent name is indicated by the longest chain i.e. a
pentane.

The name of this compound is 2,2,4,4-tetramethylpentane.

(b) $\overset{\displaystyle C_2H_5}{\underset{\displaystyle CH_3}{\overset{\displaystyle |}{\underset{\displaystyle |}{H_5C_2-C-CH_2-CH_3}}}}$ $\therefore$ $\underset{1}{CH_3}-\underset{2}{CH_2}-\underset{\underset{\displaystyle CH_3}{\overset{\displaystyle CH_2}{|}}}{\overset{\displaystyle CH_3}{\underset{3}{C}}}-\underset{4}{CH_2}-\underset{5}{CH_3}$

The name is 3-methyl-3-ethylpentane.

(c) $\underset{1}{CH_3}-\underset{2}{CH_2}-\underset{\underset{\displaystyle CH_3}{|}}{\overset{\overset{\displaystyle CH_3}{|}}{\underset{3}{C}}}-\underset{4}{CH_2}-\underset{5}{CH_3}$

The name is 3,3,-dimethylpentane.

(ii) Question 16. at the end of the chapter.

Indicate the number of, draw structural formulas for, and name

all of the isomers of C_6H_{14}.

You should first do questions of this type using molecular models

keeping in mind the definition of a structural isomer.

The first isomer is n-hexane:

$$\underset{1}{CH_3}-\underset{2}{CH_2}-\underset{3}{CH_2}-\underset{4}{CH_2}-\underset{5}{CH_2}-\underset{6}{CH_3}$$

n-hexane

Now break one bond (shorten the carbon chain to 5) and reform it on

another carbon:

$$\underset{1}{CH_3}-\underset{\underset{\displaystyle CH_3}{|}}{\underset{2}{CH}}-\underset{3}{CH_2}-\underset{4}{CH_2}-\underset{5}{CH_3}$$

2-methylpentane

and

$$\underset{1}{CH_3}-\underset{2}{CH_2}-\underset{\underset{\displaystyle CH_3}{|}}{\underset{3}{CH}}-\underset{4}{CH_2}-\underset{5}{CH_3} \qquad \text{3-methylpentane}$$

306

Now break two bonds (shorten the carbon chain to 4) and reform them on different carbons:

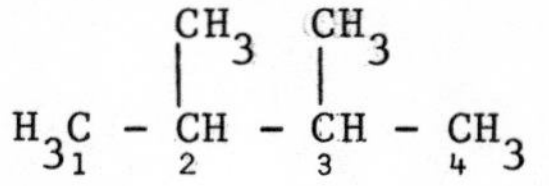

$$H_3\underset{1}{C} - \underset{2}{C}H - \underset{3}{C}H - \underset{4}{C}H_3$$

with CH_3 on carbon 2 and CH_3 on carbon 3

2,3-dimethylbutane
<u>and</u>

$$H_3\underset{1}{C} - \underset{2}{C} - \underset{3}{C}H - \underset{4}{C}H_3$$

with CH_3 above carbon 2 and $CH3$ below carbon 2

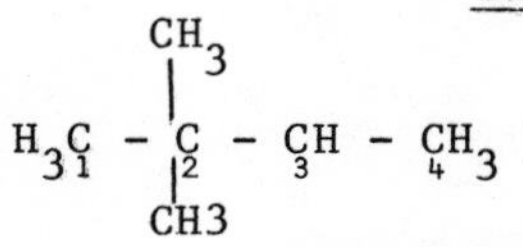

2,2-dimethylbutane

Since we cannot shorten the carbon chain beyond 4, there are <u>5</u> structural isomers.

2.(i) Question 25. at the end of the chapter.

Draw out the structural formula and give the IUPAC name for

(a) $(CH_3)_3CCH=CH_2$.

(b)   trans-$(CH_3)_2CHCH=CHCH_3$.

(c)   cis - $(CH_3)_2CHCH=CHCH_3$.

(a) 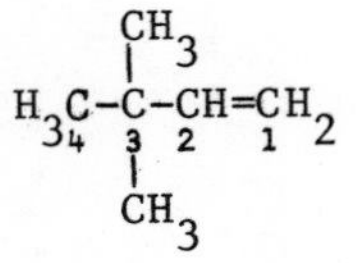
$$H_3\underset{4}{C}-\underset{3}{C}-\underset{2}{C}H=\underset{1}{C}H_2$$

with CH_3 above and CH_3 below carbon 3

Start numbering from the end closest to the double bond.

3,3,-dimethyl-1-butene

(b)   The fact that this molecule is <u>trans</u> means we should focus our attention on the double bond:

$$(CH_3)_2CH \boxed{CH=CH} CH_3$$

Using the definition of $\underline{trans}$: two substituents opposite each other, we have

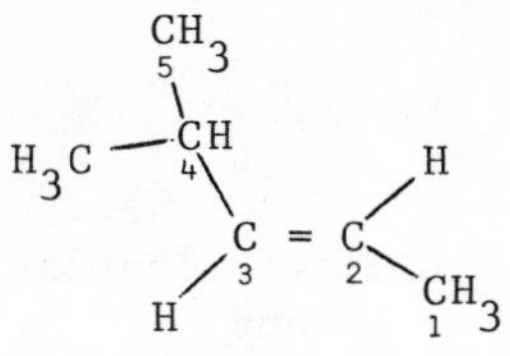

trans-4-methyl-2-pentene

(c) Again focus on the double bond and the definition of $\underline{cis}$: two substituents are adjacent.

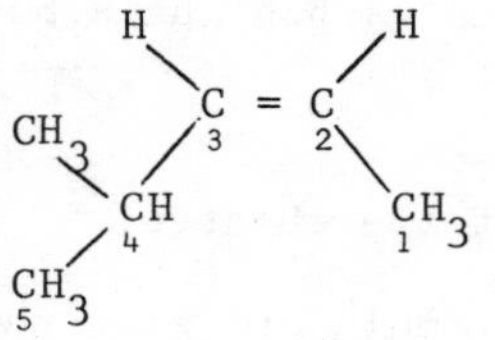

cis-4-methyl-2-pentene.

(ii) Question 30. at the end of the chapter.

Which of the following compounds can exist as cis - trans isomers? Name them.

(a) $(CH_3)_2C=CHCH_3$.

(b) $CH_3CH=CHCH_3$.

(c) $HC\equiv C(CH_3)$.

(d) $(CH_3)_2C=CCl_2$.

(a)

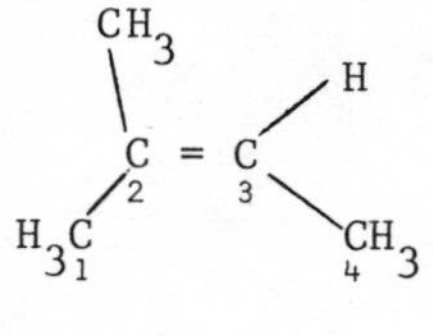

2-methyl-2-butene

No   cis - trans isomers.

(b)

H   H
 \   /
  C = C
 /3   2\
H₃C   CH₃
  4   1 3

and

H   CH₃
 \   4 3/
  C = C
 /2   3\
H₃C   H
  1

cis-2-butene   trans-2-butene

(c)   1 2 3
       H——C≡C-CH₃   No cis - trans isomers.

       1-propyne

(d)   CH₃   Cl
         \   /
          C = C   No cis - trans isomers.
         /2   1\
       H₃C   Cl
         3 3

       1,1-dichloro-2-methylpropene

3.(i) Question 38. at the end of the chapter.

   Write structural formulas for

   (a)   m-dichlorobenzene

   (b)   p-dinitrobenzene.

   (c)   o-dihydroxybenzene.

   (d)   1,2,3-tribromobenzene

   We will use the short-hand notation for benzene i.e. C_6H_6 ≡ .

(a)   (b)

(c)

(d)

(ii) Question 39. at the end of the chapter.

Name the following structures:

(a)

(b)

(a) o–dihydroxybenzene or 1,2-dihydroxybenzene.

(b) 1,3,5-trichlorobenzene.

4.(i) Question 47 at the end of the chapter.

    (a)   Propane was chlorinated and four **dichloropropane**

          products were isolated from the products. What are

          their structures?

    (b)   Each was chlorinated further. A gave one trichloro product,

          B gave two, and C and D gave three each. What are the

          structures of A and B?

(a)

$$CH_3CH_2\underset{\underset{Cl}{|}}{\overset{\overset{Cl}{|}}{CH}} \qquad\qquad CH_3-\underset{\underset{Cl}{|}}{\overset{\overset{Cl}{|}}{C}}-CH_3$$

$$CH_3\underset{\underset{}{}}{\overset{\overset{Cl}{|}}{CH}}-\overset{\overset{Cl}{|}}{CH_2} \qquad\qquad \overset{\overset{Cl}{|}}{CH_2}-CH_2-\overset{\overset{Cl}{|}}{CH_2}$$

(b) A must be $CH_3-\underset{\underset{Cl}{|}}{\overset{\overset{Cl}{|}}{C}}-CH_3$ since it is the only one that can produce the

same trichlorinated derivative by chlorination at either end (the chlorination positions are identical): i.e.

$$CH_3 - \overset{\displaystyle \overset{Cl}{|}}{\underset{\displaystyle \underset{Cl}{|}}{C}} - \overset{\displaystyle \overset{Cl}{|}}{CH_2} \cdot$$

B must be $\overset{\displaystyle \overset{Cl}{|}}{CH_2} - CH_2 - \overset{\displaystyle \overset{Cl}{|}}{CH_2}$ since it contains exactly two different sites

for chlorination: i.e.

$$\overset{\displaystyle \overset{Cl}{|}}{CH_2} - CH_2 - \underset{\displaystyle \underset{Cl}{|}}{\overset{\displaystyle \overset{Cl}{|}}{CH}} \qquad \text{and} \qquad \overset{\displaystyle \overset{Cl}{|}}{CH_2} - \overset{\displaystyle \overset{Cl}{|}}{CH} - \overset{\displaystyle \overset{Cl}{|}}{CH_2} \cdot$$

The remaining two molecules have 3 different positions available for

chlorination each.

(ii) Question 48. at the end of the chapter.

What weight of liquid oxygen would be required to completely

combust the fuel for a rocket filled with 1 liter of kerosene

(assume that the average composition of kerosene is $n\text{-}C_{14}H_{30}$

and the density is 0.90 g ml^{-1}).

The equation for combustion is

$$n\text{-}C_{14}H_{30} + \frac{43}{2}O_2 \longrightarrow 14CO_2 + 15H_2O$$

1 liter of kerosene weighs 1000 ml x 0.90 $\frac{g}{ml}$ = 900 g.

$$\# \text{ moles kerosene} = \frac{900g}{198\frac{g}{mole}} = 4.5 \text{ moles.}$$

$$4.5 \text{ mole kerosene} \times \frac{^{43}/_2 \text{ mole } O_2}{1 \text{ mole kerosene}} \times \frac{32g \text{ } O_2}{1 \text{ mole } O_2} = 3100 \text{ g } O_2 \cdot$$

311

5. (i) Question 58. at the end of the chapter.

Indicate the structures and the names of the products expected from the reaction of propene with

(a) HI.

(b) HI (if a free radical mechanism could be made to occur).

(c) Br_2.

(d) H_2SO_4.

(a) $CH_3CH=CH_2$ + HI $\xrightarrow{\text{ionic}}$ $CH_3\overset{\underset{|}{I}}{C}H-CH_3$

2-iodopropane

(b) $CH_3CH=CH_2$ + HI $\xrightarrow{\text{free radical}}$ $CH_3CH_2\overset{\underset{|}{I}}{C}H_2$

1-iodopropane

(c) $CH_3CH=CH_2$ + Br_2 $\longrightarrow$ $CH_3\overset{\underset{|}{Br}}{C}H-\overset{\underset{|}{Br}}{C}H_2$

1,2-dibromopropane

(d) $CH_3CH=CH_2$ + H_2SO_4 $\longrightarrow$ $CH_3\overset{\underset{|}{OSO_3H}}{C}H-CH_3$

isopropyl hydrogen sulfate

(ii) Question 60 at the end of the chapter.

What reagent would you add to what alkene to make

(a) 2-bromo-2-methylbutane?

(b) $CH_3CHBrCH_3$?

(c) CH_2BrCH_2Cl?

(a) We wish to make
$$H_3C - \underset{\underset{CH_3}{|}}{\overset{\overset{Br}{|}}{C}} - CH_2 - CH_3 \ .$$

Remembering Markovnikov's rule:

$$H_3C-\underset{\underset{CH_3}{|}}{C}=CH-CH_3 + HBr \longrightarrow H_3C - \underset{\underset{CH_3}{|}}{\overset{\overset{Br}{|}}{C}} - \overset{\overset{H}{|}}{CH} - CH_3$$

2-methyl-2-butene

(b) Again using Markovnikov's rule:

$$CH_2=CHCH_3 + HBr \longrightarrow CH_3 - \overset{\overset{Br}{|}}{CH}-CH_3$$

(c) $CH_2=CH_2 + Cl-Br \longrightarrow \underset{\underset{Br}{|}}{CH_2}-\underset{\underset{Cl}{|}}{CH_2}$

6.(i) Question 72 at the end of the chapter.

Starting with benzene or toluene ($C_6H_5-CH_3$) outline all steps in the synthesis of

(a) p-dichlorobenzene.

(b) 2-bromo-4-nitrotoluene (the carbon to which CH_3 is attached is labelled 1.).

(c) 4-nitro-1,2-dibromobenzene.

(d) 2-nitro-1,4-dibromobenzene.

(a)

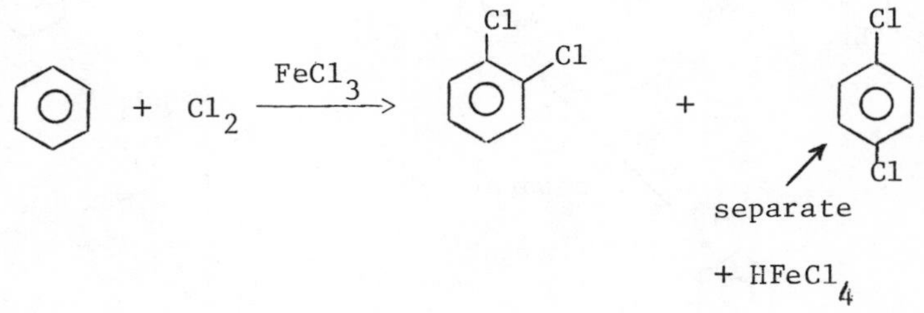

separate

+ HFeCl$_4$

(b)

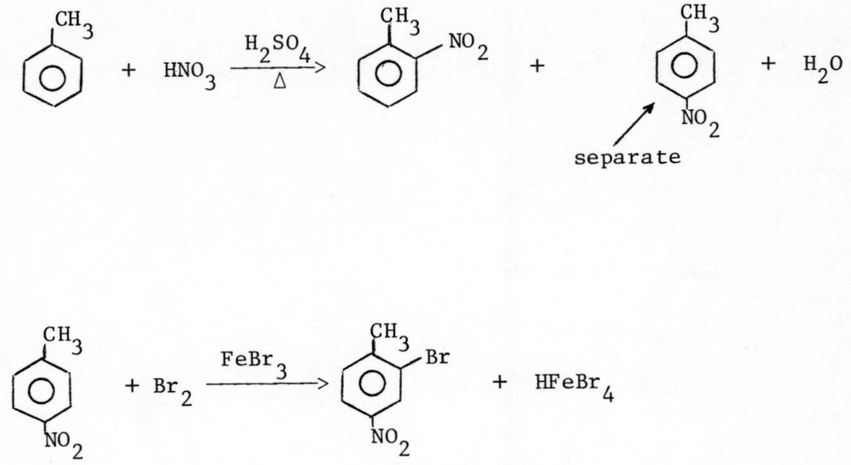

$CH_3 + HNO_3 \xrightarrow[\Delta]{H_2SO_4}$ (o-nitrotoluene) $+$ (p-nitrotoluene) $+ H_2O$

separate

(p-nitrotoluene) $+ Br_2 \xrightarrow{FeBr_3}$ (product) $+ HFeBr_4$

(c)

(benzene) $+ HNO_3 \xrightarrow[\Delta]{H_2SO_4}$ (nitrobenzene) $+ H_2O$

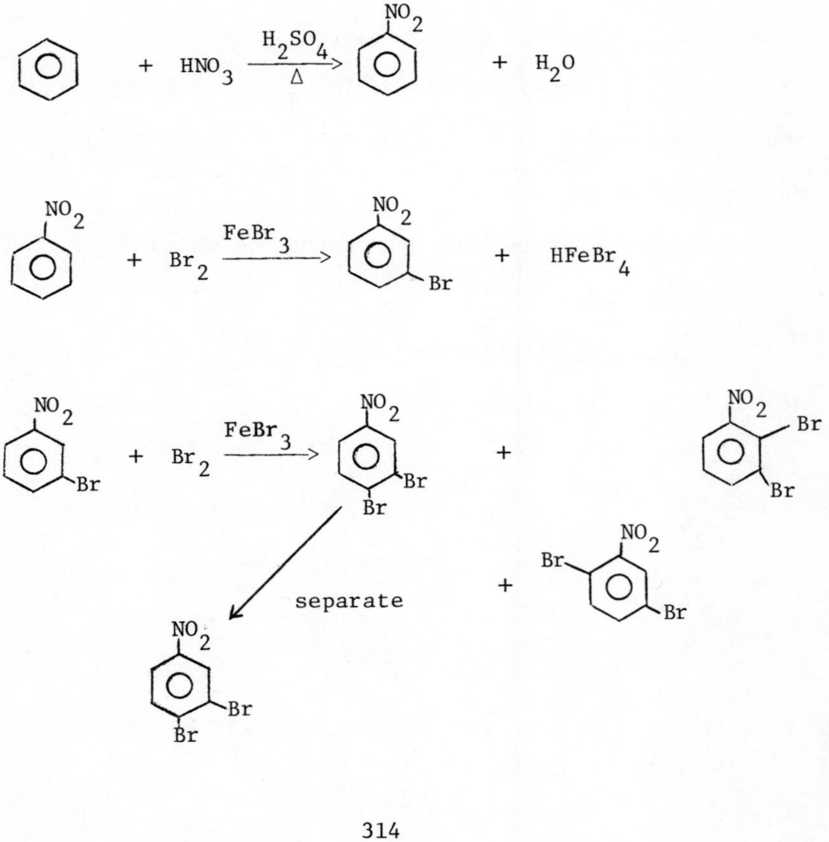

(nitrobenzene) $+ Br_2 \xrightarrow{FeBr_3}$ (m-bromonitrobenzene) $+ HFeBr_4$

(m-bromonitrobenzene) $+ Br_2 \xrightarrow{FeBr_3}$ (product) $+$ (product) $+$ (product)

separate

314

(d) Same as (c) only separate

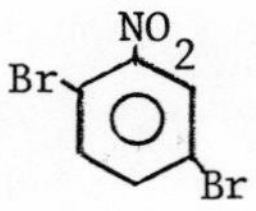

$$Br-\underset{\underset{Br}{}}{\overset{NO_2}{\bigcirc}}$$

(ii) Question 75. at the end of the chapter.

Describe how you would prepare m−nitrotoluene.

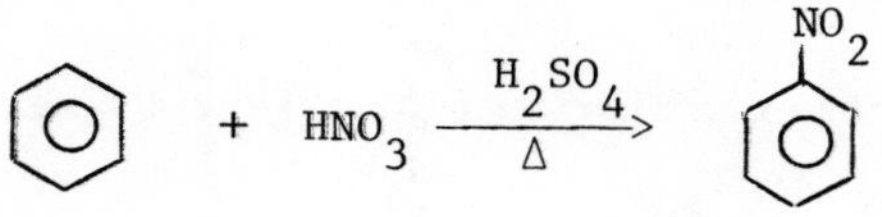

$$\bigcirc \quad + \quad HNO_3 \quad \xrightarrow[\Delta]{H_2SO_4} \quad \overset{NO_2}{\bigcirc}$$

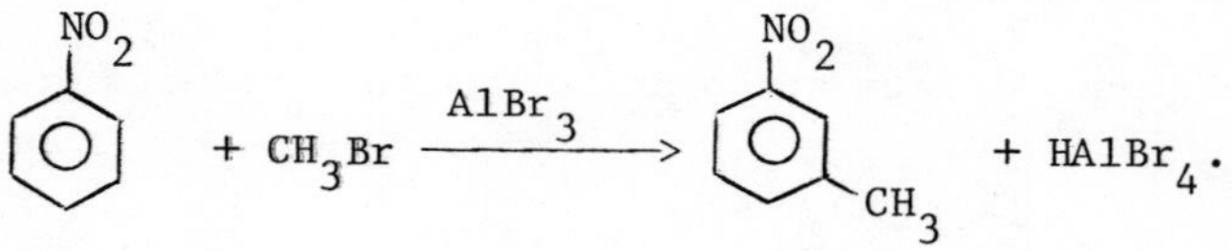

$$\overset{NO_2}{\bigcirc} \quad + \quad CH_3Br \quad \xrightarrow{AlBr_3} \quad \overset{NO_2}{\underset{CH_3}{\bigcirc}} \quad + \quad HAlBr_4 \ .$$

(iii) Question 76. at the end of the chapter.

Predict the product(s) of the following reactions:

(a) 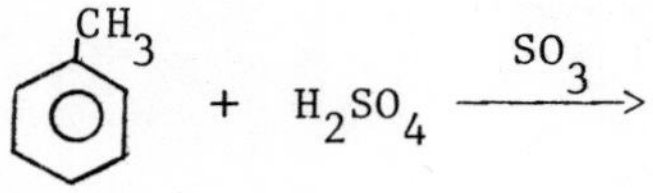 $\overset{CH_3}{\bigcirc} \quad + \quad H_2SO_4 \quad \xrightarrow{SO_3}$

(b) 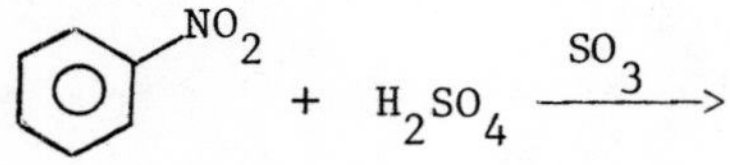 $\overset{NO_2}{\bigcirc} \quad + \quad H_2SO_4 \quad \xrightarrow{SO_3}$

(c) 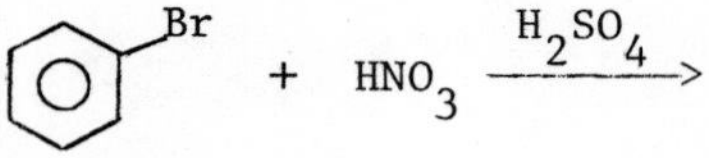 $\overset{Br}{\bigcirc} \quad + \quad HNO_3 \quad \xrightarrow{H_2SO_4}$

(a) Since CH_3 is an electron releasing substituent it favours ortho

315

and para attack in electrophilic aromatic substitution. The products are

(b)   The NO_2 group is electron withdrawing and favours meta attack. The product is

(c)   The Br group directs ortho and para: the products are:

C. Test Yourself

1. Indicate the number of, draw structural formulas for, and name all of the isomers of C_7H_{16}.

2. Distinguish between the terms saturated and unsaturated.

3. Give the structural formula of:

(a) 3-methyl-3-ethylpentane.

(b) 2,3,5-trimethyl-3-ethylhexane.

(c) 2,2,4-trimethylpentane. (isooctane).

(d) 2-bromobutane.

4. Give the structural formula of:

(a) 1,1,4-tribromo-1-butene.

(b) 2-pentene.

(c) cis-3,4-dimethyl-3-hexene.

(d) 3,6-dimethyl-1-octene.

5. Indicate which of the following compounds can exist as cis-trans isomers and name them.

(a) 4-methyl-2-pentene.

(b) 1-pentene.

(c) 2-methyl-2-butene.

(d) 2-chloro-2-butene.

6. Name the following:

(a) $HC{\equiv}C-CH(CH_3)_2$.

(b)

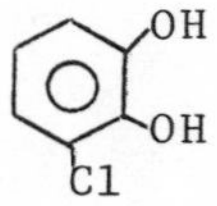

317

(c)

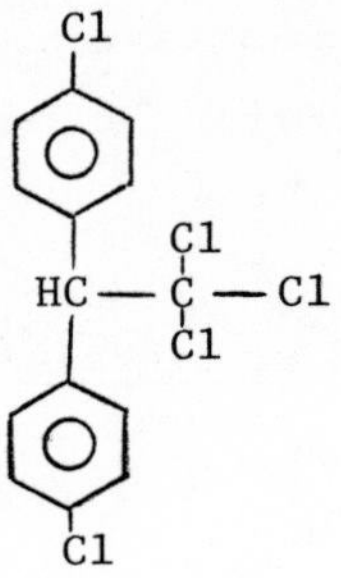

This is DDT.

(d)

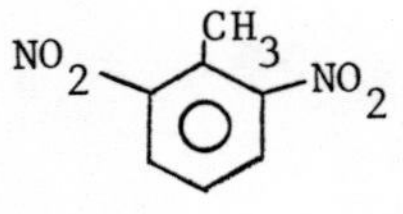

7. How many isomeric products are obtained from the monochlorination of the following compounds? Name the products.

(a) ethane

(b) propane

(c) n-butane

(d) 2-methylpropane (isobutane)

8. Write a balanced equation for the combustion of 2,2,4-trimethyl-pentane (isooctane).

9. Complete the following equations:

(a) 2-methylpropene + Br_2 $\xrightarrow{CCl_4}$

(b) CH_2=CHCl + HI $\longrightarrow$

(c) $(CH_3)_2$C=CH_2 + HBr $\xrightarrow{peroxide}$

(d) CH_3C≡CH $\xrightarrow{HCl}$ $\xrightarrow{HI}$

(e) CH_3C≡CH $\xrightarrow{Br_2}$ $\xrightarrow{Br_2}$

10. Give the structures and names of the products expected from reaction of $(CH_3)_2C=CH_2$ with:

   (a) HI

   (b) I_2

   (c) H_2SO_4

   (d) HBr (peroxide)

   (e) H_2 (Pt catalyst)

   (f) peroxide.

11. Complete the following reactions:

   (a)  ⬡ + HNO_3 $\xrightarrow{H_2SO_4}$

   (b) 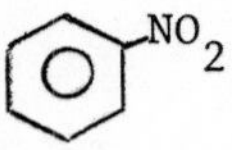 ⬡NO$_2$ + Br_2 $\xrightarrow{FeBr_3}$

   (c) 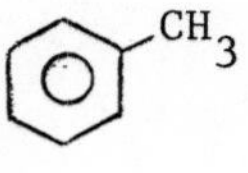 ⬡CH$_3$ + HNO_3 $\xrightarrow{H_2SO_4}$

   (d) 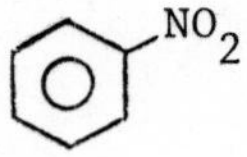 ⬡NO$_2$ + CH_3CH_2Cl $\xrightarrow{AlCl_3}$

   (e)  ⬡ + H_2SO_4 $\xrightarrow{SO_3}$

319

22

Organic Molecules Containing Some Common Functional Groups

A. <u>Points of Importance</u>.

1. Functional Groups

A <u>functional group</u> defines the structure and properties of a family of organic compounds. A number of common <u>functional group families</u>, the <u>functional group structure</u>, and an example are given below:

<u>functional group family</u>	<u>structure</u>	<u>example</u>
amine	$-N\!<$	$CH_3CH_2\!-\!N\!\!\begin{smallmatrix}H\\H\end{smallmatrix}$
alcohol	$-OH$	$CH_3\!-\!OH$
ether	$-O-$	$CH_3CH_2\!-\!O\!-\!CH_2CH_3$
alkyl halide	$-X$ (halide)	$CH_3CH_2\!-\!I$

320

aldehyde	$-C{\underset{H}{\overset{O}{\lessgtr}}}$	$CH_3-C{\underset{H}{\overset{O}{\lessgtr}}}$
ketone	$-\overset{O}{\underset{\parallel}{C}}-$	$CH_3-\overset{O}{\underset{\parallel}{C}}-CH_3$
carboxylic acid	$-\overset{O}{\underset{\parallel}{C}}-OH$	$CH_3-\overset{O}{\underset{\parallel}{C}}-OH$
ester	$-\overset{O}{\overset{\parallel}{C}}_{OR}$	$CH_3-\overset{O}{\underset{\parallel}{C}}-O-CH_3$
acid chloride	$-\overset{O}{\overset{\parallel}{C}}_{Cl}$	$CH_3-\overset{O}{\underset{\parallel}{C}}-Cl$
amide	$-\overset{O}{\overset{\parallel}{C}}-N{<}$	$H-\overset{O}{\overset{\parallel}{C}}-N{\overset{CH_3}{\underset{CH_3}{<}}}$

You should learn the typical reactions of these functional groups. Other examples of different organic compounds and their functional group(s) are given below:

cinnamaldehyde
(cinnamon)

ethyl butanoate
or
ethyl butyrate
(pineapple)

acetylsalicylic acid
(aspirin)

methyl salicylate
(oil of wintergreen)

Three other families of compounds that you will meet are the amino acids, heterocyclics, and carbohydrates. The amino acids are listed in table 22-5. Heterocylics are abundant and result when some atom other than carbon (for example, oxygen or nitrogen) is incorporated into a ring molecule. Carbohydrates are polyhydroxy aldehydes, polyhydroxy-ketones, or compounds that can be hydrolyzed to them.

2. Stereoisomerism

Stereoisomers are molecules that are different but that have identical functional groups. For example lactic acid can exist in two forms:

$$
\begin{array}{ccc}
\text{COOH} & & \text{COOH} \\
| & & | \\
\text{H} - \text{C} - \text{OH} & \quad | \quad & \text{HO} - \text{C} - \text{H} \\
| & & | \\
\text{CH}_3 & & \text{CH}_3
\end{array}
$$

These two forms cannot be superimposed no matter how you twist and turn them. The two isomers above are drawn as if there were a mirror between them. Thus enantiomers (or optical isomers) result for compounds which are mirror images but that cannot be superimposed. Prove this to yourself by building models. For the example of lactic acid above the central carbon atom is called an asymmetric carbon - a carbon that has four different groups attached. When molecules contain more than one asymmetric carbon the maximum number of different configurations is given by 2^n, where n is the number of different kinds of asymmetric carbon atoms. These configurations are not all necessarily enantiomers. (See example problem 3(i)(c).)

B.   Types of Problems

1.   Reactions and functional groups.

2.   The carbonyl functional group.

3.   Stereoisomerism.

Examples

1.   Question 11. at the end of the chapter.

Generalize from our discussion in Sec. 23-1 of halogen

derivatives, and indicate the main products expected from the

reactions of n-butyl bromide (1-bromobutane) with

   (a) KOH(aq)

   (b) KOH (alcohol)

   (c) NH_3

   (d) $NaSCH_3$

   (e) $CH_3COO^-Na^+$.

(a) $CH_3CH_2CH_2CH_2-Br + KOH(aq) \longrightarrow CH_3CH_2CH_2CH_2-OH + K^+Br^-$.

(b)   Alkyl halides can be produced by adding the hydrogen halide across a
double bond and the reverse reaction can be brought about by very
strongly basic conditions i.e. KOH (alcohol).

$CH_3CH_2CH_2CH_2-Br + KOH(alcohol) \longrightarrow CH_3CH_2CH=CH_2 + H_2O + K^+Br^-$.

(c) $CH_3CH_2CH_2CH_2-Br + NH_3 \longrightarrow CH_3CH_2CH_2CH_2-NH_3^+ + Br^-$

(d) $CH_3CH_2CH_2CH_2-Br + NaSCH_3 \longrightarrow CH_3CH_2CH_2CH_2-SCH_3 + Na^+ Br^-$

(e) $CH_3CH_2CH_2CH_2-Br + CH_3COO^-Na^+ \longrightarrow CH_3CH_2CH_2CH_2-O-\overset{O}{\overset{\|}{C}}-CH_3 + Na^+Br^-$.

2.   Question 31. at the end of the chapter.

Indicate the products of the reaction of CH_3CHO with

(a) $NaHSO_3$

(b) C_6H_5MgBr followed by hydrolysis.

(c) $KMnO_4$ (acidic)

(d) $LiAlH_4$

(e) NH_2OH

(f) $C_2H_5NHNH_2$

(g) C_2H_5OH.

(a) $CH_3C{\overset{O}{\underset{H}{\diagdown}}} + NaHSO_3 \longrightarrow CH_3C{\overset{OH}{\underset{\underset{SO_3}{|}}{\diagdown H}}}\ ^{\ominus}\ + Na^{\oplus}$

(b) $CH_3C{\overset{O}{\diagdown H}} + C_6H_5MgBr \longrightarrow CH_3C{\overset{OMgBr}{\underset{\underset{C_6H_5}{|}}{\diagup H}}} \xrightarrow{H_2O} CH_3C{\overset{OH}{\underset{\underset{C_6H_5}{|}}{\diagup H}}}$

(c) $CH_3C{\overset{O}{\diagdown H}} \xrightarrow[\text{acidic}]{KMnO_4} CH_3C{\overset{O}{\diagdown OH}}$ This is an oxidation.

(d) $CH_3C{\overset{O}{\diagdown H}} \xrightarrow{LiAlH_4} CH_3CH_2OH.$ This is a reduction.

(e) $CH_3C{\overset{O}{\diagdown H}} + NH_2OH \longrightarrow CH_3\underset{H}{C} = NOH + H_2O$

(f) $CH_3C{\overset{O}{\diagdown H}} + C_2H_5NHNH_2 \longrightarrow CH_3\underset{H}{C} = NNHC_2H_5$

(g) $CH_3C{\overset{O}{\diagdown H}} + C_2H_5OH \longrightarrow CH_3C{\overset{OC_2H_5}{\underset{\underset{OH}{|}}{\diagdown H}}}$

324

.(i) Question 53. at the end of the chapter.

Indicate the number of stereoisomers expected for

(a) CHClBrI.

(b) $CH_3CHBrCHOHCH_3$

(c) $CH_3CHBrCHBrCH_3$

(d) $CH_2OHCHOHCHOHCHOHCHO$.

a) This molecule contains one asymmetric carbon and thus there are two

stereoisomers.

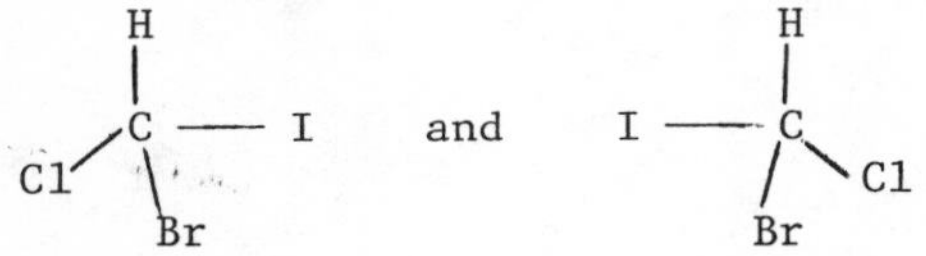

b) The structure of this molecule is

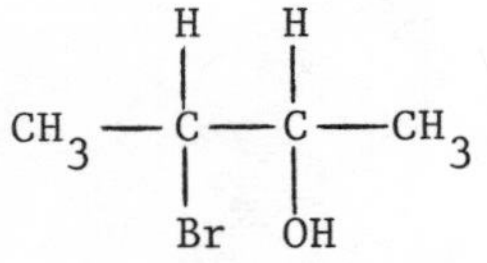

and there are <u>two</u> asymmetric carbons and
therefore 2^2 = <u>4</u> stereoisomers.

c) The structure of the molecule is

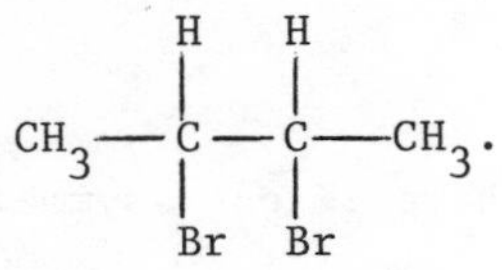

. This molecule has two asymmetric centers
but with identical groups attached. Thus

there are a total of <u>three stereoisomers</u> of which there are <u>two</u>

<u>enantiomers</u> (optically active) and one <u>meso</u> form (optically inactive).

Build models to visualize these stereoisomers!

(d) The structure is

$$H - \underset{\underset{OH}{|}}{\overset{\overset{H}{|}}{C}} - \underset{\underset{H}{|}}{\overset{\overset{OH}{|}}{\underset{*}{C}}} - \underset{\underset{H}{|}}{\overset{\overset{OH}{|}}{\underset{*}{C}}} - \underset{\underset{OH}{|}}{\overset{\overset{H}{|}}{\underset{*}{C}}} - \underset{\underset{H}{}}{\overset{\overset{}{}}{C}} \overset{O}{=\!\!\!\!}$$

.   There are three asymmetri

carbons marked with an *.

Thus there are $2^3 = \underline{8}$ stereoisomers.

(ii) Question 55. at the end of the chapter.

Which of the following can exist as enantiomers?

(a) $CH_3CH_2CHClCH_3$

(b) $CH_3CH_2CH_3$

(c) $CH_3COCH_2CH_3$

(d) $CH_3COCHClCH_3$

(e) CH_3CO_2H

(f) $CH_3CHICO_2CH_3$.

(a) The structure of the molecule is

$$CH_3 - \underset{\underset{H}{|}}{\overset{\overset{H}{|}}{C}} - \underset{\underset{Cl}{|}}{\overset{\overset{H}{|}}{\underset{*}{C}}} - CH_3$$ and there is one asymmetric carbon.

Therefore the molecule can exist as enantiomers.

(b) For $$CH_3 - \underset{\underset{H}{|}}{\overset{\overset{H}{|}}{C}} - CH_3$$ there are no asymmetric carbons and th

molecule cannot exist as enantiomers.

(c) For $CH_3 \overset{\overset{\displaystyle O}{\|}}{-C} - CH_2CH_3$ there are no asymmetric carbons

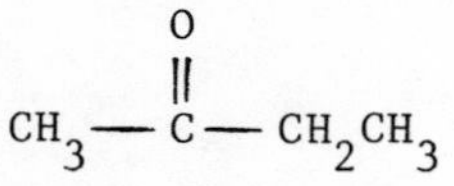

and the molecule cannot exist as enantiomers.

(d) 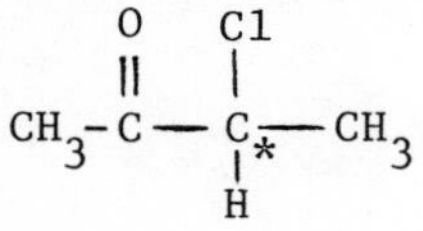 $CH_3-C \overset{\overset{\displaystyle O}{\|}}{} - \underset{\underset{\displaystyle H}{|}}{\overset{\overset{\displaystyle Cl}{|}}{C_*}} - CH_3$ can exist as enantiomers.

(e) 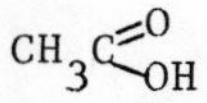 $CH_3C\overset{\nearrow O}{\underset{\searrow OH}{}}$ cannot exist as enantiomers.

(f) 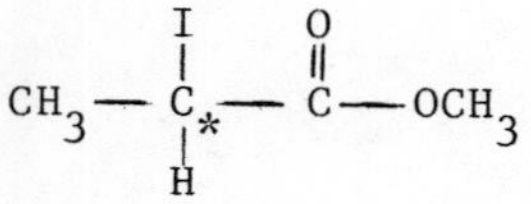 $CH_3 - \underset{\underset{\displaystyle H}{|}}{\overset{\overset{\displaystyle I}{|}}{C_*}} - \overset{\overset{\displaystyle O}{\|}}{C} - OCH_3$ can exist as enantiomers.

. As a final example let us work through <u>general summary question</u> 59.

at the end of the chapter.

Draw structural formulas to represent the principal products obtained

in the following reactions:

(a) 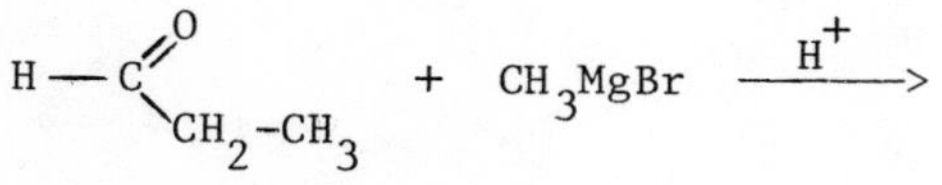 $H - C \overset{\nearrow O}{\underset{\searrow CH_2-CH_3}{}}$ + CH_3MgBr $\xrightarrow{\quad H^+ \quad}$

(b) 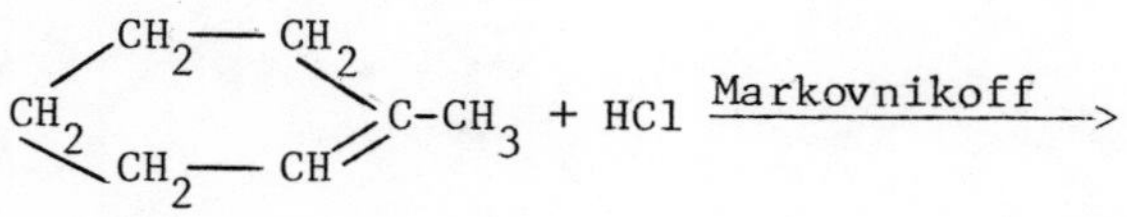 $\underset{CH_2}{\overset{CH_2—CH_2}{\diagdown}} \underset{CH_2—CH}{\overset{}{\diagup}} C\text{-}CH_3$ + HCl $\xrightarrow{\text{Markovnikoff}}$

(c) $CH_3 - \underset{\underset{\displaystyle H}{|}}{C} = C \overset{\diagup CH_3}{\underset{\diagdown CH_3}{}}$ + Br_2 $\xrightarrow[\text{ROOR}]{\text{trace}}$

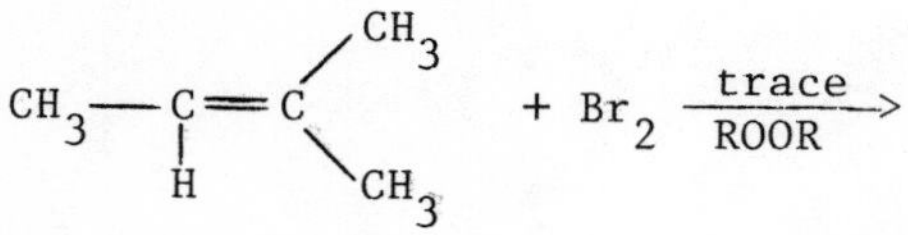

(d) 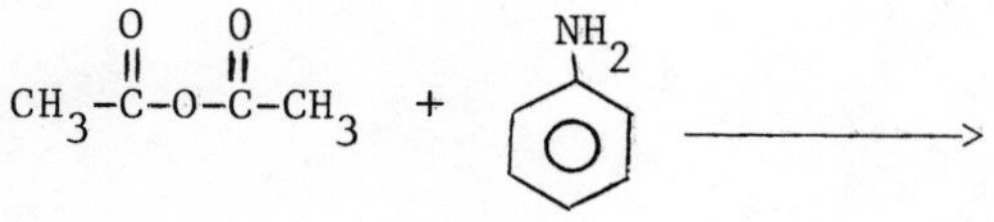 $CH_3\text{-}\overset{\overset{\displaystyle O}{\|}}{C}\text{-}O\text{-}\overset{\overset{\displaystyle O}{\|}}{C}\text{-}CH_3$ + ⟨benzene ring with NH_2⟩ $\xrightarrow{\qquad\qquad}$

(e) 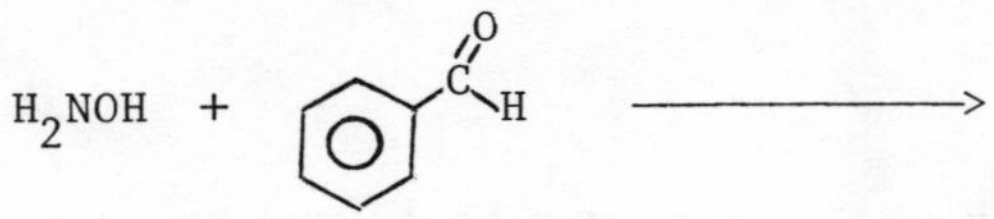 H_2NOH +

(f)

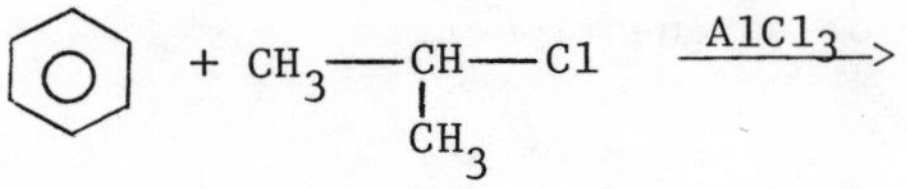

(g)

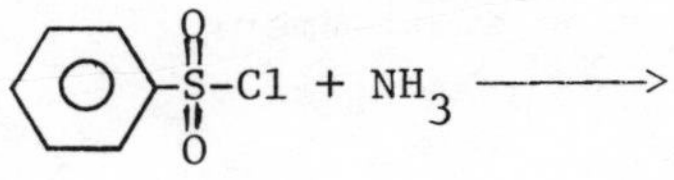

(Consider the $-SO_2Cl$ group as analogous to an acid chloride, $-COCl$.)

(a)

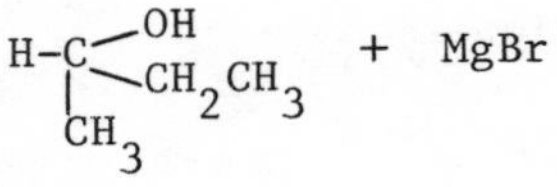

$+$ MgBr

(b)

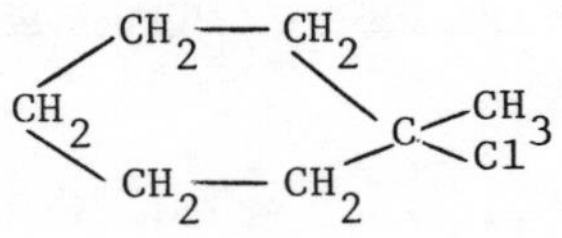

(c)

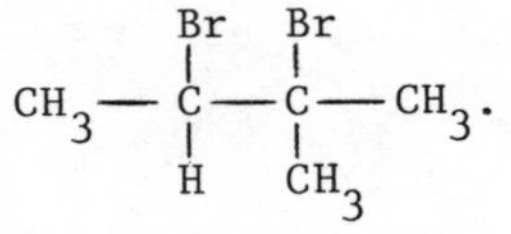

(d)

$+$ $HO\overset{O}{\overset{\|}{C}} - CH_3$.

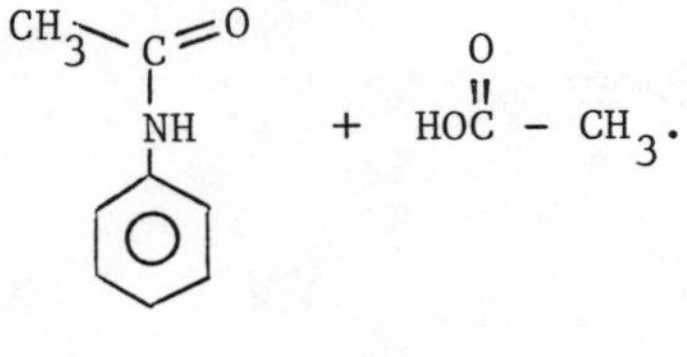

(e)

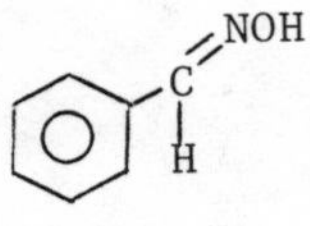

(f) 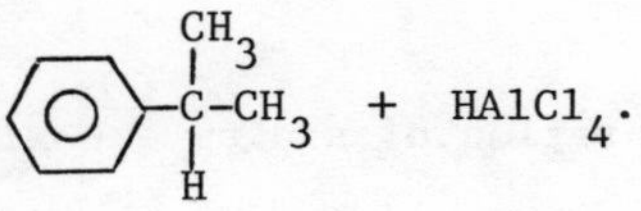

(g) 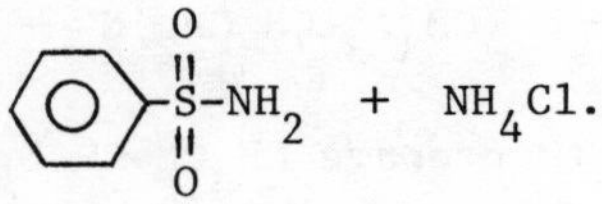

C. Test Yourself

1.(i) Write an equation for the preparation of sodium ethoxide.

(ii) Complete the following equation

$$(CH_3)_3N \ + \ \underline{\hspace{3cm}} \ \longrightarrow \ (CH_3)_3(CH_3CH_2)N^+ \ + \ I^-.$$

(iii) What alkyl halide would be used to prepare $CH_3CH=CH$ by way of an elimination reaction?

(iv) Complete the following equation.

$$+ \ H_2O \ \xrightarrow{\ H^+\ }$$

(v) Write an equation for the preparation of CH_3CONH_2 starting with an acid chloride.

(vi) The Tollen's test allows one to distinguish between _____ and _____.

(vii) Using the general reaction

$$RC\!\!-\!\!X \ + \ B^- \ \rightarrow \ RC\!\!-\!\!B \ + \ X^-$$

write equations for $X = OH^-$, $B = PCl_5$ and $X = OH^-$, $B = NH_3$. Name the products.

(viii) What type of functional group(s) is(are) found in carbohydrates

2. Complete the following statements or equations

(i) Aldehydes are easily oxidized to _____.

(ii) _____ react readily with $NaHSO_3$.

(iii) Acid chlorides react with alcohols to form _____

(iv) Alcohols can be produced by reduction of _____

(v) Ketones and aldehydes undergo _____ reactions.

(vi) $CH_3C\overset{O}{\underset{H}{\diagup}}$ + $NH_2NHCONH_2$ ——————>

(vii) ⬡$-C\overset{O}{\underset{H}{\diagup}}$ + CN^- $\xrightarrow{H^+}$

(viii) ⬡$-C\overset{O}{\underset{H}{\diagup}}$ + $2Ag(NH_3)_2^+$ $\xrightarrow{OH^-}$

(ix) ⬡$-\overset{O}{\overset{\|}{C}}-CH_3$ + $2Ag(NH_3)_2^+$ $\xrightarrow{OH^-}$

3. For the following compounds (a) indicate the number of stereoisomers
   and (b) indicate which compounds are capable of existing as enantiomers.

(i) CHClBrI

(ii) $Co(OH_2)_2Cl_2$ (tetrahedral)

(iii) $CHCl_2Br$

(iv) $Co(OH_2)(NH_3)ClBr$ (tetrahedral)

(v) $Pt(NH_3)_2Cl_2$ (planar)

(vi) $Pt(NH_3)(OH_2)IBr$ (planar)

4. How could you determine if the metal complex $[M(CN)(Br)(I)(NH_3)]$ was
   planar or tetrahedral? (Assume possible stereoisomers can be isolated.)

5. How many stereoisomers and how many optically active isomers exist for
   the following?

(i) $CH_3CHOHCHClCH_3$

(ii) $CH_3CH_2CHClCHClCH_3$

331

(iii) $CH_3CHClCHClCH_3$

(iv) $CH_3CHOHCH_2OH$.

6.   Distinguish between the following using examples:
     stereoisomers,   enantiomers,   and   diastereoisomers.

23

Biochemistry

A. Points of Importance

1. Proteins

Polypeptide: a molecule derived from two or more amino acids

linked by peptide bonds.

Protein: a biologically active polypeptide containing many amino

acids. Usually the molecular weight is $\geq$ 10,000.

Peptide bond: the bond between amino acid units in a protein or poly-

peptide; $\begin{smallmatrix} H & O \\ | & \| \\ -N- & C- \end{smallmatrix}$ Water is eliminated when a peptide bond is formed.

Primary structure: the specific order of the amino acids.

Secondary structure: the way the protein chain coils as a result of

hydrogen bonding, electrostatic links, or disulfide links.

Tertiary structure: the overall bulk shape of the protein.

Quaternary structure: the arrangement of tertiary units for proteins

with more than one chain.

333

Since proteins are made up of amino acids joined by peptide bonds there must be two distinct amino acids at each end of the peptide chain. The N-terminal group contains a free amino group while the C-terminal group contains a free carboxyl group. End group analysis involves the reaction of different reagents specifically with either of the two ends, hydrolysis, and identification of the reagent-end group molecule. For example, the tripeptide glycylalanylphenylalanine reacts with the N-terminal specific reagent 2,4-dinitrofluorobenzene according to the equation below:

$$\overset{+}{H_3}NCH_2\overset{\overset{O}{\|}}{C}-NH-\underset{\underset{CH_3}{|}}{CH}-\overset{\overset{O}{\|}}{C}-NH-\underset{\underset{CH_2C_6H_5}{|}}{CH}-COO^- \quad + \quad O_2N-\langle\bigcirc\rangle\overset{-F}{\underset{-NO_2}{}}$$

$$\downarrow \text{ base}$$

$$O_2N-\langle\bigcirc\rangle\underset{NO_2}{}-NH-CH_2-\overset{\overset{O}{\|}}{C}-NH-\underset{\underset{CH_3}{|}}{CH}-\overset{\overset{O}{\|}}{C}-NH-\underset{\underset{CH_2C_6H_5}{|}}{CH}-COO^-$$

The product is then "totally" hydrolyzed by aq. HCl and heat to give

$$O_2N-\langle\bigcirc\rangle\underset{NO_2}{}-NH-CH_2-COOH \quad +\overset{+}{H_3}N-\underset{\underset{CH_3}{|}}{CH}-COOH \quad + \overset{+}{H_3}N-\underset{\underset{CH_2C_6H_5}{|}}{CH}-COOH$$

The N-terminal amino acid can be identified since it is labelled by the 2,4-dinitrophenyl group.

2. Metabolism and Energy Storage

   Some of the energy produced by the breakdown of foods is stored for use by the cell in compounds such as adenosine triphosphate. The energy is released by formation of adenosine diphosphate by hydrolysis:

334

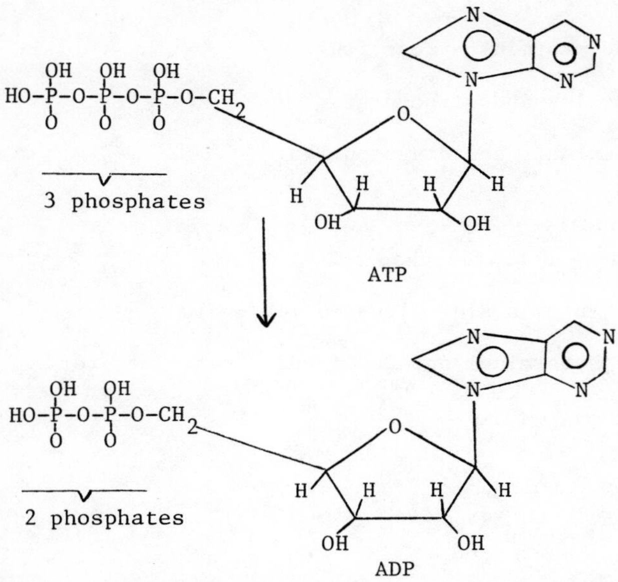

ATP

ADP

A common reaction for ATP to undergo is phosphorylation:

$$ATP + R–OH \rightarrow ADP + R–OPO_3H_2 + energy.$$

335

B.   Types of Problems

   1.   Amino Acids and proteins - reactions.

   2.   Biochemical cycles and structures.

   3.   Effect of pollutants and drugs on man.

Examples

1.(i) Question 4. at the end of the chapter.

   What is the difference in ala - leu and leu - ala?

   By convention the N-terminal end is listed first. Therefore

   ala - leu has the structure

$$H_3\overset{+}{N}-\underset{\underset{CH_3}{|}}{CH}-\overset{\overset{O}{\|}}{C}-NH-\underset{\underset{\underset{CH(CH_3)_2}{|}}{CH_2}}{CH}-COO^-$$

   The structure of leu - ala is then

$$H_3\overset{+}{N}-\underset{\underset{\underset{CH(CH_3)_2}{|}}{\underset{CH_2}{|}}}{CH}-\overset{\overset{O}{\|}}{C}-NH-\underset{\underset{CH_3}{|}}{CH}-COO^-$$

(ii) Question 9. at the end of the chapter.

   An octapeptide upon complete hydrolysis yielded ala, glu, gly, leu, met, val, and two molecules of cys. After partial hydrolysis, the fragments gly - cys, leu - val - gly, cys - glu, cys - ala, cys - glu - met, and met - cys were obtained. What is the sequence in the octapeptide?

      To solve this problem we overlap the small peptides so as to give a consistent sequence.

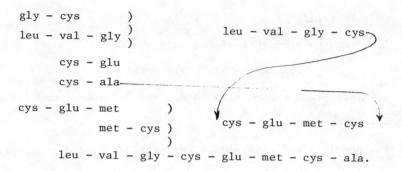

gly - cys   )
leu - val - gly)
)
       cys - glu
       cys - ala ─────────

cys - glu - met   )
           met - cys)
)
leu - val - gly - cys - glu - met - cys - ala.

leu - val - gly - cys

cys - glu - met - cys

2. Question 14. at the end of the chapter.

What are the two half-reactions in the Krebs cycle when NAD^+ converts

malic acid

$$HOOCCH_2-\underset{\underset{OH}{|}}{\overset{\overset{H}{|}}{C}}-COOH \qquad \text{to oxaloacetic acid}$$

$$HOOCCH_2-\overset{\overset{O}{\|}}{C}-COOH?$$

$$HOOC-CH_2-\underset{\underset{OH}{|}}{\overset{\overset{H}{|}}{C}}-COOH \xrightarrow{-[2H]} HOOC-CH_2-\underset{\underset{O}{\|}}{C}-COOH$$

$$NAD^+ \xrightarrow{+[2H]} NADH + H^+$$

3. Question 26. at the end of the chapter.

What does an antihistamine do to relieve hay fever?

    All antihistamines block the release of histamine,

$$\underset{NH}{\overset{N \quad CH_2CH_2NH_2}{\bigcirc}} \quad ,$$

which causes allergy symptoms.

C.   Test Yourself

    1. Draw the structure of the polypeptide

                    ser – met – gly – ala – phe

    indicating the peptide bonds, the N-terminal end and the

    C-terminal end.

    2. Complete hydrolysis of a hexapeptide yielded arg, tyr, val, phe,

    cys. After partial hydrolysis, the fragments tyr – cys, arg – phe,

    cys, val – arg, and phe – tyr were obtained.

    (a) What are the possible sequences in the hexapeptide?

    (b) How can you elminate one of the possibilities from part (a)?

    (c) Draw the structure.

    3. Complete the following equations:

    (a) 2,4-dinitrofluorobenzene + cys – ala $\longrightarrow$ A

    (b) A $\xrightarrow{\text{HCl}}$
$\quad$heat

    (c) 2,4-dinitrofluorobenzene + lys – met – tyr $\longrightarrow$ B

    (d) B $\xrightarrow{\text{HCl}}$
$\quad$heat

4. What characteristic reactions might you expect a small polypeptide to

undergo at the C-terminal end?

5. (a) Write an equation to describe the production of energy in our

       bodies by the oxidation of food, $C_6H_{12}O_6$(s).

    (b) In what two ways is this energy channeled?

    (c) How is the energy in ATP made available for chemical reactions?

6. Represent the adenylic unit of RNA (a nucleotide) clearly showing the

nucleoside and the heterocyclic base.

7. The liver can detoxify compounds by oxidation as shown for a
    "martini" below:

$$CH_3CH_2OH \xrightarrow[\text{enzymes}]{\text{liver}} CH_3C\overset{O}{\underset{H}{\diagup\!\!\!\diagdown}} \xrightarrow{[O]} CH_3COOH$$

The acetic acid is then oxidized to CO_2 and water.

(a) Write the "detoxification" reaction for methyl alcohol
     (wood alcohol).

(b) Is the product of the reaction desired?

(c) Can you suggest an antidote?

8. How do sulfa drugs prevent bacterial infection?

9. Explain the danger of drinking alcohol while taking large amounts of
    headache preparations containing caffeine.

Answers to Section C – Test Yourself

<u>Chapter 1</u>*

1.(i) $238.03 \frac{g}{m} \times \frac{1m}{6.023 \times 10^{23} \text{ atoms}} = 3.952 \times 10^{-22} \frac{g}{\text{atom}}$.

(ii) $1.152 \times 10^{-23} \frac{g}{\text{atom}}$.

(iii) $291.26 \frac{g}{m} \times \frac{1m}{6.023 \times 10^{23} \text{ molecules}} = 4.836 \times 10^{-22} \frac{g}{\text{molecule}}$

     (Assume average atomic weights).

(iv) $25.0 \text{ g} \times \frac{1 \text{ mole}}{386.66 \text{ g}} \times \frac{6.023 \times 10^{23} \text{ molecules}}{1 \text{ mole}} = 3.89 \times 10^{22} \text{ molecules}$.

(v) $28.0 \text{ g} \times \frac{1 \text{ mole}}{371.47 \text{ g}} \times \frac{6.023 \times 10^{23} \text{ molecules}}{1 \text{ mole}} \times \frac{8 \text{ atoms carbon}}{1 \text{ molecule}} =$

     $= 3.63 \times 10^{23} \text{ atoms}$.

(vi) $88.0 \text{ g} \times \frac{1 \text{ mole}}{230.66 \text{ g}} \times \frac{24.02 \text{ g C}}{1 \text{ mole}} = 9.16 \text{ gC}$.

* m = mole.

(vii) $454 \text{ g} \times \dfrac{1 \text{ mole}}{354.49 \text{ g}} = 1.28 \text{ moles.}$

2. Average atomic mass = $23.98504 \times 0.7870 + 24.98584 \times 0.1013$

$+ 25.98259 \times 0.1117$

$= 24.31.$

3. $1.0080 = x(1.0078) + (1-x)(2.0141)$
   $x = 0.9998$

99.98% of isotope – 1.0078, $^{1}_{1}\text{H}$

0.02% of isotope – 2.0141, $^{2}_{1}\text{H}$ or Deuterium.

4.

Species	# protons	# electrons	# neutrons
$^{32}_{16}\text{S}$	16	16	16
$^{79}_{34}\text{Se}^{2-}$	34	36	45
$^{56}_{26}\text{Fe}^{3+}$	26	23	30
$^{40}_{20}\text{Ca}^{2+}$	20	18	20
$^{27}_{13}\text{Al}^{3+}$	13	10	14
$^{19}_{9}\text{F}^{1-}$	9	10	10

Chapter 2.

1. $CoCl_2H_{12}O_6$.

2. $K_2Cr_2O_7$.

3. % C = 49.45, % H = 8.30, % N = 12.82, % O = 29.43: $C_9H_{18}N_2O_4$.

4. 5.00×10^3 mole $CuFeS_2 \times \dfrac{4 \text{ moleSO}_2}{2 \text{ moleCuFeS}_2} = 10.0 \times 10^3$ moleSO_2.

5. 5.00×10^3.

* 6. moles $SiO_2 = \dfrac{25.0 \text{ g}}{60.1 \frac{\text{g}}{\text{m}}} = 0.416$ m.

0.416 m SiO_2 x $\dfrac{1mP_4}{6mSiO_2} = 6.93 \times 10^{-2}$ mP_4

mass $P_4 = 6.93 \times 10^{-2}$ m x 123.88 $\dfrac{\text{g}}{\text{m}} = 8.58$ g.

7. moles $BaSO_4 = \dfrac{0.1632 \text{ g}}{233.40 \frac{\text{g}}{\text{mole}}} = 6.992 \times 10^{-4}$ moles = moles $BaCl_2$ present.

mass $BaCl_2 = 6.992 \times 10^{-4}$ m x 208.24 $\dfrac{\text{g}}{\text{m}} = 0.1456$ g.

% $BaCl_2 = \dfrac{0.1456 \text{ g}}{0.525 \text{ g}}$ x 100 = 27.7%.

8. (i) SiO_2

(ii) $Ca_3(PO_4)_2$.

9(i) moles $KMnO_4 = \dfrac{20.0 \text{ g}}{158.04 \frac{\text{g}}{\text{m}}} = 0.127$ m; moles $H_2O_2 = \dfrac{10.0 \text{ g}}{34.01 \frac{\text{g}}{\text{m}}} = 0.294$ m

moles HCl = $\dfrac{1.0 \times 10^2 \text{ g}}{36.45 \frac{\text{g}}{\text{m}}} = 2.74$ m. **Divide** number of moles of each species

by its coefficient in the equation. The smallest number obtained gives the limiting leagent i.e. H_2O_2.

0.294 m H_2O_2 x $\dfrac{2mMnCl_2}{5mH_2O_2}$ x $\dfrac{125.84 \text{ g}}{1mMnCl_2} = 14.8$ g $MnCl_2$.

(ii) 66.4%.

10. 1mA x $\dfrac{5mC}{3mA}$ x $\dfrac{3mE}{2mC}$ x $\dfrac{6mP}{1mE} = 15mP$.

  equation   1   2   3

* m = mole.

Chapter 3

*1. (i)   moles = $5.0\ell \times 6.0 \frac{m}{\ell} = 30.m$

molarity = $\frac{30.m}{15\ \ell} = 2.0M$

(ii)   moles $Ca(NO_3)_2$ = moles Ca^{2+} = $4.0\ell \times 0.20 \frac{m}{\ell} = 0.80m$

moles $CaCl_2$ = moles Ca^{2+} = $6.0\ell \times 0.80 \frac{m}{\ell} = 4.8m$

total moles Ca^{2+} = 5.6m and $[Ca^{2+}] = \frac{5.6m}{10.\ell} = 0.56M.$

(iii)   moles H_2SO_4 = $0.250\ell \times 0.15 \frac{m}{\ell} = 0.038m$

$\frac{0.038m}{0.025\frac{m}{\ell}} = 1.5\ell.$

(iv) $\frac{mass\ HCl}{ml} = \frac{37gHCl}{100g} \times \frac{1.19g}{ml} = 0.440g \frac{HCl}{ml}$

$0.440 \frac{g\ HCl}{ml} \times \frac{1mHCl}{36.45gHCl} \times \frac{1000ml}{\ell} = 12.1M.$

(v)   moles $NaClO_4$ = $0.225\ell \times 1.65 \frac{m}{\ell} = 0.371m$

mass $NaClO4$ + $0.371m \times 122.45\frac{g}{m} = 45.4g.$

2. (i)   moles CuS = $\frac{25.0g}{95.60\frac{g}{m}} = 0.262m. \rightarrow \div 3 = 0.0873$

moles HNO_3 = $0.140\ell \times 2.00\frac{m}{\ell} = 0.280m \rightarrow \div 8 = 0.035$

The limiting reagent is HNO_3.

* m = mole.

$* \ 0.280mHNO_3 \ \times \ \dfrac{3mCu(NO_3)_2}{8mHNO_3} \ = \ 0.105mCu(NO_3)_2.$

$mass \ Cu(NO_3)_2 \ = \ 0.105m \ \times \ 187.54\dfrac{g}{m} \ = \ 19.7g.$

(ii) $require \ 0.262mCuS \ \times \ \dfrac{8mHNO_3}{3mCuS} \ = \ 0.699mHNO_3.$

$volume \ = \ \dfrac{0.699m}{2.00m/\ell} \ = \ 0.350\ell \ or \ 350m\ell.$

3. $moles \ Cr_2O_7{}^{2-} \ = \ 0.0152 \ \ell \ \times \ 0.200\dfrac{m}{\ell} \ = \ 0.00304 \ moles.$

$0.00304 \ moles \ Cr_2O_7{}^{2-} \ \times \ \dfrac{6 \ moles \ Fe^{2+}}{2 \ moles \ Cr_2O_7{}^{2-}} \ = \ 0.00912 \ moles \ Fe^{2+}.$

$mass \ iron(II) \ fumarate \ = \ 0.00912 \ moles \ \times \ \dfrac{169.9g}{mole} \ = \ 1.55g \ total.$

There are 15.5 mg iron(II) fumarate in each tablet.

$Percent \ iron(II) \ fumarate \ in \ each \ tablet \ = \ \dfrac{0.0155 \ g}{0.386 \ g} \ \times \ 100 \ = \ 4.02\%.$

4. $moles \ NaOH \ = \ 0.0148 \ \ell \ \times \ 0.100 \ \dfrac{m}{\ell} \ = \ 0.00148m$

$0.00148mNaOH \ \times \ \dfrac{1mH_2SO_4}{2mNaOH} \ \times \ \dfrac{1mSO_3}{1mH_2SO_4} \ \times \ \dfrac{2mSO_2}{2mSO_3} \ = \ 0.000740mSO_2.$

$mass \ SO_2 \ = \ 0.000740m \ \times \ 64.06\dfrac{g}{m} \ = \ 0.0474gSO_2.$

$concentration \ = \ \dfrac{0.0474g}{200\ell} \ = \ 2.37 \ \times \ 10^{-4} \ g\ell^{-1}.$

5. $moles \ excess \ HCl \ = \ 0.100\ell \ \times \ 1.000\dfrac{m}{\ell} \ = \ 0.100mHCl$

$moles \ NaOH \ used \ = \ 0.0320\ell \ \times \ 1.000\dfrac{m}{\ell} \ = \ 0.0320 \ mNaOH \ = \ moles \ HCl \ left$

$moles \ HCl \ that \ reacted \ with \ CaCO_3 \ = \ 0.100m \ -0.0320m \ = \ 0.0680mHCl.$

$0.0680mHCl \ \times \ \dfrac{1mCaCO_3}{2mHCl} \ \times \ \dfrac{100.1gCaCO_3}{moleCaCO_3} \ = \ 3.40gCaCO_3.$

* m = mole.

344

$$\text{Percent purity} = \frac{3.40g}{5.00g} \times 100 = 68.0\%.$$

*6. moles NaOH = $0.050\ell \times 0.184\dfrac{m}{\ell}$ = 0.0092 moles

moles $HClO_4$ = $0.050\ell \times 0.085\dfrac{m}{\ell}$ = 0.0043 moles.

moles NaOH left after reaction = 0.0092 −0.043 = 0.0049m.

$$[Na^+] = \frac{0.0092m}{0.500\ell} = 0.018 \quad ; \quad [ClO_4^-] = \frac{0.0043m}{0.500\ell} = 0.0086$$

$$[OH^-] = \frac{0.0049m}{0.500\ell} = 0.0098 \quad ; \quad [H^+] = \frac{Kw}{[OH^-]} = \frac{1.0 \times 10^{-14}}{0.0098} = 1.0 \times 10^{-12}.$$

Chapter 4.

1.(i)   Heat was put into the system and since the container was open to the atmosphere w = o and ΔE = +q.

   (ii)   As the container cools heat is given off and work done on the system since the container collapses. Therefore ΔE = −q + w.

2.(i) $C(graphite) + 2H_2(g) + \frac{1}{2}O_2(g) \rightarrow CH_3OH(\ell)$.

   (ii) $\frac{1}{2}N_2(g) + \frac{1}{2}O_2(g) \rightarrow NO(g)$

   (iii) $H_2(g) + \frac{1}{8}S_8(s) + 2O_2(g) \rightarrow H_2SO_4(\ell)$.

   (iv) $C(graphite) \rightarrow C(diamond)$.

   (v) $Ca(s) + C(graphite) + \frac{3}{2}O_2(g) \rightarrow CaCO_3(s)$.

3.   See text and study guide.

4.(i) $E = \dfrac{hc}{\lambda} = \dfrac{6.63 \times 10^{-27}\text{erg-sec} \times 3.00 \times 10^{10}\text{cm s}^{-1}}{3303 \times 10^{-8}\text{ cm}} = 6.02 \times 10^{-12}$ ergs.

6.02×10^{-12} ergs $\times \dfrac{1 \text{ kcal}}{4.184 \times 10^7 \text{ergs}} = 1.44 \times 10^{-19}$ kcal.

* m = mole.

(ii) $\nu = \dfrac{c}{\lambda} = \dfrac{3.00 \times 10^{10} \text{cm s}^{-1}}{3303 \times 10^{-8} \text{ cm}} = 9.08 \times 10^{14} \text{s}^{-1}.$

5. $E = h\nu$ and $\nu = \dfrac{E}{h} = \dfrac{7.4 \times 10^{-12} \text{erg}}{6.63 \times 10^{-27} \text{erg-sec}} = 1.1 \times 10^{15} \text{s}^{-1}.$

6. $E = \dfrac{hc}{\lambda} = \dfrac{6.63 \times 10^{-27} \text{erg-sec} \times 3.00 \times 10^{10} \text{cm s}^{-1}}{26.9 \times 10^{-8} \text{cm}} = 7.39 \times 10^{-10} \text{erg}.$

Chapter 5.

1.(i)   (d)   N

(ii)   (a) Cl^-

(iii)   (d)   5

(iv)   (e)   Co

(v)   (b) $n = 4, \ell = 0, m_\ell = 0, m_s = \frac{1}{2}.$

(vi)   (d)   P

(vii)   (b)   Be

(viii) (b)   90°

(ix)   (d)   Cs

(x)   (e)   KF

2.(i) $[Ar]3d^8 4s^2$

(ii) $[Ar]3d^3$

(iii) $[Ar]3d^8$

(iv) $[Ar]$

(v) $[Ar]3d^{10} 4s^2 4p^5$

3.(i)

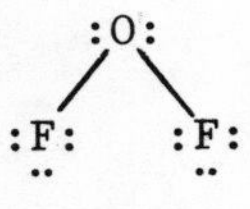

:O:

:F: :F:

(ii)

$$\left[\begin{array}{c} H \\ H-N-H \\ H \end{array} \right]^+$$

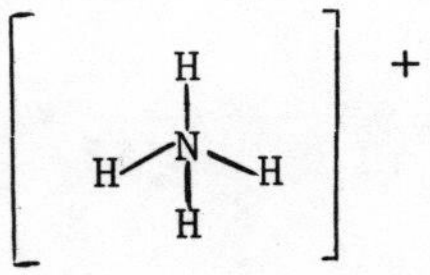

(iii)

$$\begin{array}{ccc}
& H & \\
& | & \\
& :\overset{..}{O}: & \\
& | & \\
:\overset{..}{O} & N & \overset{..}{O}: \\
\end{array}
\quad\longleftrightarrow\quad
\begin{array}{ccc}
& H & \\
& | & \\
& :\overset{..}{O}: & \\
& | & \\
:\overset{..}{O}: & N & \overset{..}{O}: \\
\end{array}$$

(iv) $\quad [\overset{..}{N} = C = \overset{..}{O}]^{-} \;\longleftrightarrow\; [:N \equiv C - \overset{..}{O}:]^{-} \;\longleftrightarrow\; [:\overset{..}{N} - C \equiv O:]^{-}$

(v) $\left[\begin{array}{c} :\overset{..}{O}: \\ | \\ S: \\ \diagup \;\; \diagdown \\ :\overset{..}{O} \quad \overset{..}{O}: \end{array} \right]^{2-}$
    (vi) $\quad [\overset{..}{N} = N = \overset{..}{N}]^{-}$

(vii) $\left[\begin{array}{c} \overset{..}{N} \\ \diagup\;\diagdown \\ :\overset{..}{O} \quad \overset{..}{O}: \end{array} \right]^{-} \;\longleftrightarrow\; \left[\begin{array}{c} \overset{..}{N} \\ \diagup\;\diagdown \\ :\overset{..}{O}: \quad \overset{..}{O}: \end{array} \right]^{-}$

(viii) $\begin{array}{c} \overset{..}{P} \\ \diagup\;|\;\diagdown \\ :\overset{..}{Cl} \quad \overset{..}{Cl}: \\ :\overset{..}{Cl}: \end{array}$
    (ix) $\quad \overset{..}{S} = C = \overset{..}{S}$
    (x) $\begin{array}{c} H \diagdown \quad \diagup H \\ C = C \\ H \diagup \quad \diagdown H \end{array}$

4. (i) (a) S^{2-}, Cl^{-}, K^{+}, Ca^{2+} etc.

(b) Zn^{2+}, Ga^{3+}, Cu^{1+}.

(c) Li^{+}, Be^{2+}, H^{-}.

(ii) $1s^{2}2d^{1}2p^{1}$: impossible.

$1s^{2}2s^{2}$: ground state Be.

$1s^{2}2s^{3}$: impossible.

$1s^{2}2s^{2}2p^{2}$: ground state C.

$1s^{2}2s^{2}2p^{6}2d^{1}$: impossible.

$1s^{2}2s^{2}2p^{3}$: ground state N.

$1s^{2}2s^{1}3s^{1}$: excited state B.

347

5. (a) Ge > Si > C.   (b) Li > Be > B.

    (c) S^{2-} > Cl^- > K^+ > Ca^{2+}.

Chapter 6.

1. $\dfrac{CaO}{BaS}$ = 1.33, $\dfrac{BaO}{CaS}$ = 1.03.

2.(i) $\ddot{O} = C = \ddot{O}$, linear, 180°, sp hybridisation, 2σ bonds and 2π bonds, no dipole moment.

(ii)

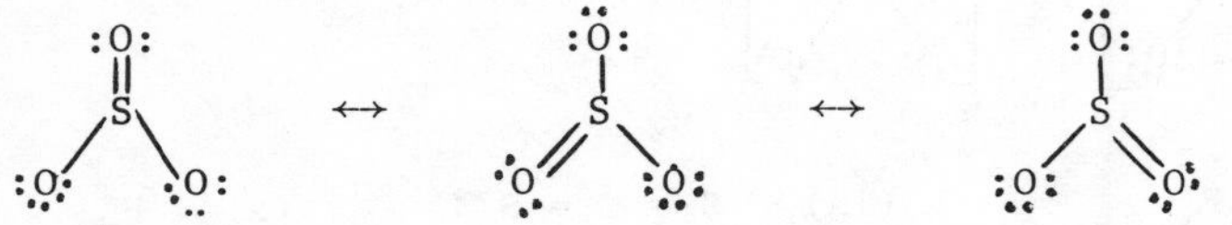

planar with all O-S-O angles 120°, sp^2 hybridisation, 3σ bonds and 1π bond, no dipole moment.

(iii)

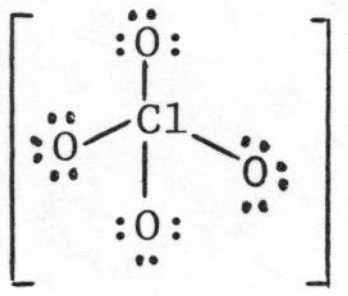

, tetrahedral, 109°, sp^3 hybridisation, 4σ bonds, no dipole moment.

(iv)

, trigonal planar with angle approximately 120°, sp^2 hybridisation, 3σ bonds and 1π bond, dipole moment.

(v)

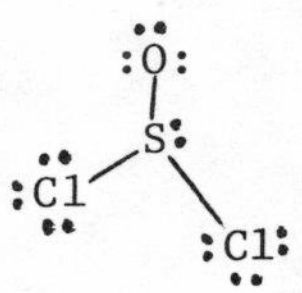

, trigonal pyramidal, O-S-Cl and Cl-S-Cl angles approximately 109°, sp^3 hybridisation, 3σ bonds, dipole moment.

348

(vi)

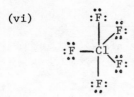 , trigonal bipyramidal, two angles at 90° and three at 120°, sp^3d hybridisation, 5σ bonds, no dipole moment.

3. (i)

Species	Bond Order	Prediction
O_2^+	2.5	exists
Li_2^+	0.5	possibly exists
Li_2^{2+}	0	does not exist
Ne_2	0	does not exist
B_2	1.0	exists
Be_2^+	0.5	possibly exists
NO^-	2.0	exists.

(ii)

Species	Bond Order	
NO	2.5	bond lengths increase → $NO^+ < NO < NO^-$.
NO^-	2.0	dissociation energies increase →
NO^+	3.0	$NO^- < NO < NO^+$.

Chapter 7.

1. $V_2 = \dfrac{P_1\,T_2\,V_1}{P_2\,T_1} = \dfrac{(760\ \text{mmHg})\,(148°K)\,(1.85\ \ell)}{(0.0100\ \text{mmHg})\,(273°K)} = 7.62 \times 10^4\ \ell.$

2. $P_2 = \dfrac{P_1\,V_1\,T_2}{V_2\,T_1} = \dfrac{(50.0\ \text{atm})\,(0.500\ \ell)\,(298°K)}{(10.0\ )\,(248°K)} = 3.00\ \text{atm}.$

3. $V = \dfrac{nRT}{P} = \dfrac{454}{16.0}$ mole $\times\ 0.0821\ \dfrac{\ell.\text{atm}}{\text{mole}°K} \times \dfrac{298°K}{1\ \text{atm}} = 694\ \ell.$

4. Develop the formula $GMW = g\,\dfrac{RT}{PV}$ and

$$GMW = 4.00 \text{ g} \times 0.0821 \frac{\ell.atm}{mole°K} \times \frac{398°K}{2.12 \ \ell} \times \frac{1}{\dfrac{740 \text{ mmHg}}{760 \text{ mmHg atm}^{-1}}}$$

$$= \frac{4.00 \times 0.0821 \times 398 \times 760}{2.12 \times 740} = 63.3 \text{ g mole}^{-1}.$$

5. $R = \dfrac{PV}{nT} = \dfrac{1.013 \times 10^5 \text{ Nm}^{-2} \times 22.4 \text{ dm}^3}{1 \text{ mole} \times 273°K} = 8.31 \times 10^3 \text{ Nm}^{-2}\text{dm}^3\text{mole}^{-1}°K^{-1}.$

6. Total moles of gas formed = 8 moles

$$V = \frac{nRT}{P} = \frac{8 \times 0.0821 \times 298}{1} = 196 \ \ell$$

7. $n_{Ag} = \dfrac{1.0}{107.9} = 9.3 \times 10^{-3}$

$$9.3 \times 10^{-3} \text{moleAg} \times \frac{1\text{mole O}_3}{6 \text{ moleAg}} \times \frac{48\text{g O}_3}{1 \text{ mole O}_3} \times \frac{\ell \text{ air}}{1.0 \times 10^{-3}\text{g O}_3} = 74 \ \ell \text{ air.}$$

8.(i) $TiBr_4$. Larger London forces.

(ii) Ba. Metallic forces stronger than London.

(iii) Br_2O. London forces plus dipole-dipole are greater than London forces alone in Br_2.

(iv) CaO. Ionic forces stronger than London plus dipole-dipole in CO.

(v) HI. Both have dipole-dipole but HI has much larger London forces.

(vi) H_2O. Both have London dispersion forces which are similar but H_2O has much larger dipole-dipole forces.

(vii) HCl. Ar and HCl have London forces but HCl also must overcome dipole-dipole forces.

9.(i) $CHCl_3 > CH_2Cl_2 > CH_3Cl$ (London + dipole-dipole forces).

(ii) Kr > Ar > Ne. (London dispersion forces).

10. $\Delta H_{sub} = \Delta H_{vap} + \Delta H_{fus} = 7.2 + 2.8 = 10. \text{ kcal mole}^{-1}.$

Chapter 8.

1.(i)   Brönsted acid: $H_2O + NH_3 \rightarrow OH^- + NH_4^+$.

        Brönsted base: $H_2O + HCl \rightarrow H_3O^+ + Cl^-$.

        Lewis acid: $H_3N: + H_2O \rightarrow H_3N-H-O-H$.

        Lewis base: $CrCl_3 + 6H_2O \rightarrow Cr(OH_2)_6^{3+} + 3Cl^-$

(ii)   Lewis acid: $BF_3 + NH_3 \rightarrow H_3N-BF_3$.

(iii)   Brönsted acid: $Al(OH_2)_6^{3+} + H_2O \rightarrow (OH_2)_5Al(OH)^{2+} + H_3O^+$.

(iv)   Brönsted Base: $C_5H_5N + H_2O \rightarrow C_5H_5NH^+ + OH^-$.

        Lewis base: $C_5H_5N + H^+ \rightarrow C_5H_5NH^+$.

2.(i) $(CH_3)_3N: + H_2O \rightarrow (CH_3)_3NH^+ + OH^-$

(ii) $CN^- + H_2O \rightarrow HCN + OH^-$

(iii) $F^- + H_2O \rightarrow HF + OH^-$.

3. $HCO_3^- + H_2O \rightarrow H_2CO_3 + OH^-$

    $HF + H_2O \rightarrow H_3O^+ + F^-$

4. $HCl + OH_2 \rightleftarrows H_3O^+ + Cl^-$ Since K_a for this reaction is very large there is little tendency of Cl^- to react with H_3O^+ or in other words to act as a base. Therefore the conjugate base, Cl^-, of the strong acid HCl is a very weak base i.e. the reaction $Cl^- + H_2O \rightleftarrows HCl + OH^-$ has a negligible K_h. On the other hand for a weak acid, say HCN, the reaction $HCN + H_2O \rightleftarrows H_3O^+ + CN^-$ proceeds to a very small extent (small K_a) and therefore CN^- will cause hydrolysis of H_2O: $CN^- + H_2O \rightleftarrows HCN + OH^-$ to a greater extent.

5.   A Brönsted conjugate base and acid are related to each other by the loss or gain of a proton. For example, in the reaction $H_2O + NH_3 \rightleftarrows OH^- + NH_4^+$ the conjugate base-acid pairs are NH_3/NH_4^+ and OH^-/H_2O.
H_2O/OH^-; H_3O^+/H_2O; OH^-/O^{2-}; H_2CO_3/HCO_3^-; HCO_3^-/CO_3^{2-}.
All are listed as acid /base pairs.

6. Both $HClO_4$ and HCl are strong acids while CH_3CO_2H, HCN, and HNO_2 are weak acids with K_a's 1.8×10^{-5}, 4.0×10^{-10}, and 4.5×10^{-4} respectively. Therefore the order of decreasing strength of the conjugate bases is

$$CN^- > CH_3CO_2^- > NO_2^- > Cl^-, ClO_4^-.$$

Chapter 9.

1.(a)   Acidic: $H_2PO_4^- + H_2O \rightleftarrows HPO_4^{2-} + H_3O^+.$

(b)   Acidic: $Fe(OH_2)_6^{2+} + H_2O \rightleftarrows (OH_2)_5Fe(OH)^{1+} + H_3O^+.$

(c)   Neutral: $NH_4^+ + H_2O \rightleftarrows NH_3 + H_3O^+$ $K_{h1} = 5.6 \times 10^{-10}.$

$$CH_3CO_2^- + H_2O \rightleftarrows CH_3CO_2H + OH^- \quad K_{h2} = 5.6 \times 10^{-10}$$

(d)   Neutral.

(e)   Basic: $HS^- + H_2O \rightleftarrows H_2S + OH^-.$

2.   The label implies that large concentrations of NH_4^+ and OH^- ions can exist in solution. This is not true since NH_3 is a weak base. See p. 120.

3.(a) $Ba^{2+} + 6H_2O \rightarrow Ba(OH_2)_6^{2+}$ hydration.

(b) $Fe^{2+} + 6H_2O \rightarrow Fe(OH_2)_6^{2+}$ hydration.

$Fe(OH_2)_6^{2+} + H_2O \rightleftarrows (OH_2)_5Fe(OH)^{1+} + H_3O^+$ hydrolysis.

(c) $Al^{3+} + 6H_2O \rightarrow Al(OH_2)_6^{3+}$ hydration.

$Al(OH_2)_6^{3+} + H_2O \rightleftarrows (OH_2)_5Al(OH)^{2+} + H_3O^+$ hydrolysis.

4. $K_h = \dfrac{[HF]\,[OH^-]}{[F^-]}$ and $K = \dfrac{[F^-]}{[HF][OH^-]}$. $K = K_h^{-1}.$

5. $\Delta P = X_2 P_1^o = (\dfrac{0.0781}{1.36}) \times 97.0 = 5.57$ mmHg.

6. $\Delta T = k \times$ molality $= (2.53°C\ molal^{-1}) \times 0.781\ molal = 1.98°C.$

The boiling point of the solution is b.p. (pure benzene) + ΔT

i.e. $80.1°C + 1.98°C = 82.1°C$.

7.   molality $= \dfrac{1.59}{5.12} = 0.311$ and $\dfrac{1.50\text{g caffeine}}{25.0\text{g benzene}}$ x 1000g benzene

$$= 60.0\text{g caffeine}.$$

$\therefore \dfrac{60.0\text{g caffeine}}{1000\text{g benzene}} \times \dfrac{1000\text{g benzene}}{0.311 \text{ moles caffeine}} = 193 \dfrac{\text{g}}{\text{mole}}$.

8.   See text p. 310.

9. $C_5H_5NHCl - BH^+$ and equation 9-3.

   HCN - HX and equation 9-2.

   KOH - MOH and equation 8-28.

   $HClO_4$ - MOH and equation 8-29.

   KF - C^+X^- and equation 9-6 and 9-8.

   C_5H_5N - neutral base and equation 9-4.

   $Fe(ClO_4)_2 - C^+X^-$ and equation 9-9 and 9-5.

Chapter 10.

1.(i) Rb_2S, CoS, CaS.

 (ii) SeO_2, SO_2, Cl_2O, Rb_2O.

(iii)(a) $Cs(OH_2)_6^+ + Cl^-$.

   (b) $Cr(OH_2)_6^{3+} + 3ClO_4^-$ and $Cr(OH_2)_6^{3+} + H_2O \rightleftharpoons (OH_2)_5CrOH^{2+} + H_3O^+$.

   (c) $Be(OH_2)_4^{2+} + 2Cl^-$ and $Be(OH_2)_4^{2+} + H_2O \rightleftharpoons (OH_2)_3BeOH^+ + H_3O^+$.

 (iv) (a) $LiH + H_2O \rightarrow [Li-H---H-OH] \rightarrow H_2 + LiOH$

   (b) $BeH_2 + NH_3 \rightarrow [HBe-H---H-NH_2] \rightarrow H_2 + HBeNH_2$

     $NH_2BeH + NH_3 \rightarrow NH_2Be-H---H-NH_2] \rightarrow H_2 + Be(NH_2)_2$.

   (c) $N_2O_5 + H_2O \rightarrow 2HNO_3$.

   (d) $NCl_3 + 3H_2O \rightarrow NH_3 + 3HOCl$.

(e) $BiCl_3 + H_2O \rightarrow [Cl_3Bi\text{-}OH_2] \rightarrow BiOCl + 2H^+ + 2Cl^-$.

2.   (a) $\underline{S}$, + 6.   (b) $\underline{S}$, + 4,   (c) $\underline{I}$, + 7.

3. (a) $5H_2O_2 + 2MnO_4^- + 6H^+ \rightarrow 5O_2 + 2Mn^{2+} + 8H_2O$.

   (b) $H_2O + 3Na_2S_2O_3 + 8KMnO_4 \rightarrow 3Na_2SO_4 + 3K_2SO_4 + 8MnO_2 + 2KOH$.

   (c) $3CN^- + IO_3^- \rightarrow 3CNO^- + I^-$.

4. $Cr_2O_7^{2-} + 8H^+ + 3NO_2^- \rightarrow 2Cr^{3+} + 3NO_3^- + 4H_2O$.

5. (a) N_2, 0; N_2H_2, -1; N_2H_4, -2; NH_3, -3.

   (b) $8H^+ + N_2 + 8e^- \rightarrow 2NH_3 + H_2$.

Chapter 11.

1. $\Delta H°_{rxn} = 2\,(21.60) = 43.20$ kcal mole^{-1} and $\Delta n = 0$

      $\Delta E° = \Delta H° - \Delta nRT$

          $= 43.20$ kcal mole^{-1}.

2. (a) $C_8H_{18}(\ell) + 12\frac{1}{2}O_2(g) \rightarrow 8CO_2(g) + 9H_2O(g)$

   (b) $Pb(s) + C(graphite) + \frac{3}{2}O_2(g) \rightarrow PbCO_3(s)$.

   (c) $C(graphite) + O_2(g) \rightarrow CO_2(g)$.

3. $\Delta H°_{rxn} = 2(-94.1) - [2(-26.4) + 2(21.6)]$
          $= -188.6$ kcal mole^{-1}.

4. $4NH_3(g) + 5O_2(g) \rightarrow 4NO(g) + 6H_2O(\ell)$ $-1.169 \times 10^3 kJ$

$\underline{6H_2O(\ell) + 2N_2(g) \rightarrow 4NH_3(g) + 3O_2(g)}$ $+1.530 \times 10^3 kJ$

add $2N_2(g) + 2O_2(g) \rightarrow 4NO(g)$ $0.361 \times 10^3 kJ$

$\div 4$ $\frac{1}{2}N_2(g) + \frac{1}{2}O_2(g) \rightarrow NO(g)$ $90.3 kJ.$

5. $3Pb(s) + 4CO_2(g) \rightarrow Pb_3O_4(s) + 4CO(g)$ $+397.3 kJ$

$4CO(g) \rightarrow 4C(s) + 2O_2(g)$ $4(+110.5)kJ$

$\underline{4C(s) + 4O_2(g) \rightarrow 4CO_2(g)}$ $4(-393.5)kJ$

add $3Pb(s) + 2O_2(g) \rightarrow Pb_3O_4(s)$ $-734.7 kJ$

6. $H_2(g) + Br_2(g) \rightarrow 2HBr(g)$ $-103 kJ$

$2H(g) \rightarrow H_2(g)$ $-436 kJ$

$\underline{2Br(g) \rightarrow Br_2(g)}$ $-193 kJ$

$2H(g) + 2Br(g) \rightarrow 2HBr(g)$ $-732 kJ$

$H(g) + Br(g) \rightarrow HBr(g)$ $-366 kJ$

7. $C_8H_{18}(\ell) + 12\frac{1}{2}O_2(g) \xrightarrow{\text{I}} 8CO_2(g) + 9H_2O(g)$

$\downarrow$ 9.22 kcal   II

$C_8H_{18}(g) + 12\frac{1}{2}O_2(g)$

II $= 16(-194) + 18(-110) + 12.5(118) + 7(80) + 18(99)$

$= -1267$ kcal

I = 9.22 + II = 9.22 - 1267 = -1258 kcal.

8. $Li^+(g)$ + $I^-(g)$

$\uparrow IE$ $\uparrow EA$ $\searrow \Delta H_L$

$Li(g)$ $I(g)$

$\uparrow \Delta H_s$ $\uparrow \frac{1}{2}\Delta H_D$

$Li(s)$ + $\frac{1}{2}I_2(g) \xrightarrow{\Delta H_f^\circ} Li^+I^-(s)$

$\Delta H_f^\circ = \Delta H_s + IE + \frac{1}{2}\Delta H_D + EA + \Delta H_L$

$-64.6 = 38.6 + 124 + \frac{1}{2} \times 36 - 71 + \Delta H_L$

$\Delta H_L = -174.2$ kcal mole^{-1}.

9. $\Delta H^\circ_{rxn} = [2\Delta H_f^\circ (Fe_2O_3(s)) + 8\Delta H_f^\circ (SO_2(g))] - [4\Delta H_f^\circ (FeS_2(s))]$

$-791 = 2\Delta H_f^\circ (Fe_2O_3(s)) + 8(-71.0) - 4(-42.5)$

$2\Delta H_f^\circ (Fe_2O_3(s)) = -791 + 568 - 170$

$= -393$ kcal mole^{-1}

$\Delta H_f^\circ (Fe_2O_3(s)) = -197$ kcal mole^{-1}.

Chapter 12

1.   A   +   3B $\rightleftharpoons$ 2C

   initial   1.00   3.00   0

   equilibrium 1.00 $- \frac{1}{2} \times$ 0.980   3.00 $- \frac{3}{2} \times$ 0.980   0.980

              = 0.510   = 1.53

$$K = \frac{(0.980)^2}{(0.510)(1.53)^3} = 0.526 \; M^{-2}$$

356

2.(a) right

(b) right

(c) no change (number of gaseous reactants equals number of
gaseous products).

3.

$n_{HClO_4} = n_{H^+} = 0.200 \; l \times 0.30 \; \frac{m}{l} = 0.060 \; m$

$n_{Ca(OH)_2} = 0.300 \times 0.050 = 0.015 \; m$

$n_{OH^-} = 2 \times n_{Ca(OH)_2} = 0.030 \; m.$

$H^+ + OH^- \rightarrow H_2O \; \therefore \; n_{H^+} = 0.060 - 0.030 = 0.030 \; m.$

$[H^+] = \dfrac{0.030 \; m}{0.500 \; l} = 0.060 \; M.$

pH = 1.22.

4.
$$HF + H_2O \rightleftarrows H_3O^+ + F^-$$

initial   0.10

equilibrium   0.10-x   x   x

$$7.2 \times 10^{-4} = \frac{x^2}{0.10-x}$$

$$x = [H^+] = 8.1 \times 10^{-3}$$

$$pH = 2.09.$$

5.
$$F^- + H_2O \rightleftarrows HF + OH^-$$

initial   0.10

equilibrium   0.10-x   x   x

* m = mole.

357

$$K_h = \frac{Kw}{Ka} = \frac{1.0 \times 10^{-14}}{7.2 \times 10^{-14}} = \frac{x^2}{0.10-x}$$

$$x = [OH^-] = 1.2 \times 10^{-6} \quad \text{and} \quad pH = 8.08.$$

* 6. This is a buffer problem. Since pH required is 9.00 we have $[H^+] = 10^{-9}$ and $[OH^-] = 10^{-5}$.

	NH_3	+	H_2O	$\rightleftarrows$	NH_4^+	+	OH^-
initial	0.500				$C_{NH_4^+}$		
equilibrium	$0.500-10^{-5}$				$C_{NH_4^+} + 10^{-5}$		10^{-5}

$$K_b = 1.8 \times 10^{-5} = \frac{(C_{NH_4^+})(10^{-5})}{0.500}$$

$$C_{NH_4^+} = 0.90M$$

$$n_{NH_4^+} = 0.90 \frac{m}{\ell} \times 0.500 \; \ell = 0.45 \text{ moles}$$

weight $NH_4Cl = 0.45 \; m \times 53.5 \frac{g}{m} = 24 \; g.$

7. Since the concentrations of the weak base and strong acid are the sa[me] equal volumes must be added to reach the equivalence point. Therefor[e] we have at the equivalence point

$$C_2H_5NH_2 + H^+ \longrightarrow C_2H_5NH_3^+$$

$$C_{C_2H_5NH_3^+} = \frac{0.240}{2} = 0.120M \text{ (volume is doubled).}$$

* m = mole.

$$C_2H_5NH_3^+ + H_2O \rightleftharpoons C_2H_5NH_2 + H_3O^+$$

initial   0.120

equilibrium   0.120-x   x   x

$$K_h = \frac{K_w}{K_b} = \frac{1.0 \times 10^{-14}}{5.6 \times 10^{-4}} = \frac{x^2}{0.120-x}$$

$$x = [H^+] = 1.5 \times 10^{-6}$$

$$pH = 5.82.$$

8.
$$HC_4H_6ClO_2 + H_2O \rightleftharpoons C_4H_6ClO_2^- + H_3O^+$$

initial   C

equilibrium $C - 10^{-3}$ 10^{-3} 10^{-3}

$$8.9 \times 10^{-5} = \frac{(10^{-3})^2}{C - 10^{-3}}$$

$$C = 0.012 \text{ M.}$$

$$n_{HC_4H_6ClO_2} = 0.012 \frac{mole}{\ell} \times 5.00 = 0.060 \quad \therefore \text{ mass } = 7.4 \text{ g.}$$

9.
$$HC_4H_7O_2 + H_2O \rightleftharpoons H_3O^+ + C_4H_7O_2^-$$

initial   0.240   0.120

equilibrium $0.240 - 3.00 \times 10^{-5}$ 3.00×10^{-5} $0.120 + 3.00 \times 10^{-5}$

$$K_a = \frac{(3.00 \times 10^{-5})(0.120)}{(0.240)} \qquad \text{(Neglect } 3.00 \times 10^{-5} \text{ w.r.t.}$$
$$\qquad\qquad\qquad\qquad\qquad\qquad 0.240 \text{ and } 0.120).$$

$$K_a = 1.50 \times 10^{-5}.$$

10.
$$n_{NaOH} = n_{OH^-} = \frac{4.00}{40.0} = 0.100 \text{ moles}; \quad n_{HOCN} = 0.500 \times 1.0 = 0.50$$

$$OH^- + HOCN \rightarrow H_2O + OCN^-$$

initial   0.100   0.50

equilibrium   0.40   0.10

$$C_{HOCN} = \frac{0.40}{0.500} = 0.80 \text{ M and } C_{OCN^-} = \frac{0.10}{0.500} = 0.20 \text{ M}$$

Since we have a weak acid and its conjugate base we have a buffer.

$$HOCN + H_2O \rightleftarrows H_3O^+ + OCN^-$$

initial   0.80   0.20

equilibrium 0.80-x x 0.20 + x

$$2.2 \times 10^{-4} = \frac{x(0.20)}{0.80} \qquad \text{(Neglect } x \text{ w.r.t. } 0.80 \text{ and } 0.20 \text{ M.)}$$

$$x = [H^+] = 8.8 \times 10^{-4}$$

$$pH = 3.06.$$

11. (a) $H^+ + NH_3 \rightarrow NH_4^+$

    (b) $OH^- + C_5H_5NH^+ \rightarrow C_5H_5N + H_2O$

CHAPTER 13

1. $Pb_3(PO_4)_2 \rightleftarrows 3Pb^{2+} + 2PO_4^{3-}$

$$(3x)^3(2x)^2 = 1 \times 10^{-42}$$

$$108 \, x^5 = 1 \times 10^{-42}$$

$$x^5 = 9 \times 10^{-45}$$

Use logs to find x: $\log (9 \times 10^{-45}) = -45 + 0.9542$

$$\frac{1}{5}(-45 + 0.9542) = (-9 + 0.1908)$$

$$\text{antilog } (-9 + 0.1908) = 1.6 \times 10^{-9}$$

$$\therefore \; x = 2 \times 10^{-9} \text{ (significant figures).}$$

2. $(x) (2x)^2 = 2 \times 10^{-6}$

$\qquad 4 x^3 = 2 \times 10^{-6}$

$\qquad x = 8 \times 10^{-3}$

3. Ion Product $= [K^+] [ClO_4^-] = (0.10) (0.05) = 5 \times 10^{-3}$

Since K_{sp} is greater than ion product no precipitation occurs.

4. $CaF_2(s) \rightleftharpoons Ca^{2+} + 2F^-$

$\qquad K_{sp} = [Ca^+] [F^-]^2 = (2 \times 10^{-5}) (2 \times 2 \times 10^{-5})^2 = 3 \times 10^{-14}$

5. $BaSO_4(s) \rightleftharpoons Ba^{2+} + SO_4^{2-}$

initial $\qquad 0 \qquad 0.010M$

equilibrium $\quad x \qquad x + 0.010$

$\qquad K_{sp} = 1.5 \times 10^{-9} = [Ba^{2+}] [SO_4^{2-}] = (x) (x + 0.010)$

$\qquad\qquad\qquad x = 1.5 \times 10^{-7}$

Therefore 1.5×10^{-7} moles of $BaSO_4(s)$ will dissolve.

6. Need the concentrations of Ag^+ to cause precipitation of each solid.

$Ag_2C_2O_4(s) \rightleftharpoons 2Ag^+ + C_2O_4^{2-}$

$\qquad 5.3 \times 10^{-12} = [Ag^+]^2 [0.050]$

$\qquad\qquad [Ag^+] = 1.0 \times 10^{-5}$

$\quad AgCl(s) \rightleftharpoons Ag^+ + Cl^-$

$\qquad 1.6 \times 10^{-10} = [Ag^+] [0.050]$

$\qquad\qquad [Ag^+] = 3.2 \times 10^{-9}$

Therefore $AgCl(s)$ will precipitate first.

7. IP = $[2.50 \times 10^{-5}]$ $[2.50 \times 10^{-5}]$ = 6.3×10^{-10}

Since K_{sp} is less than IP precipitation will occur.

8. $PbCO_3(s) \rightleftarrows Pb^{2+} + CO_3^{2-}$

$1 \times 10^{-13} = x^2$

$x = 3 \times 10^{-7} M$ = solubility of $PbCO_3(s)$

$Ag_2C_2O_4(s) \rightleftarrows 2Ag^+ + C_2O_4^{2-}$

$5.3 \times 10^{-12} = [2x]^2[x]$

$x = 1.1 \times 10^{-4} M$ = solubility of $Ag_2C_2O_4(s)$.

CHAPTER 14

1.(a) $\Delta G° = [2(-94.3) + 0] - [2(+20.72) + 2(-32.8)]$

$= -164.4$ kcal mole^{-1}

(b) Yes

(c) Negative. There is more disorder in the reactants than in the products.

$\Delta S° = [2(51.1) + 45.7] - [2(50.3) + 2(47.3)]$

$= -47.3$ cal mole^{-1}deg^{-1}.

(d) $\Delta G° = \Delta H° - T\Delta S°$

At 25°C we have

$\Delta H° = -164.4 + (298)(-47.3) \times 10^{-3}$ kcal mole^{-1}

$= -178.5$ kcal mole^{-1}.

2.(a) Plot log K. versus T^{-1}(°K) and obtain $\Delta H°$ from the slope and $\Delta S°$ from the intercept. See section 14-5 in text. $\Delta H° = -127.2$ kcal mole^{-1}, $\Delta S° = +32.0$ cal mole^{-1}deg^{-1}

(b) $\Delta G° = -RT\ell nK$

$= -1.987 \times 298 \times 2.303 \log (1.9 \times 10^{100})$

$$= -1363.7 \times 100.279$$

$$= -136.7 \text{ kcal mole}^{-1}$$

3.(a) $\Delta G° = \Delta H° - T\Delta S°$

$$= -10.0 - 298 \times (-5.2 \times 10^{-3})$$

$$= -10.0 + 1.5$$

$$= -8.5 \text{ kcal mole}^{-1}$$

The reaction is feasible at room temperature.

(b) The extent of the reaction will decrease at higher temperatures since $\Delta G°$ will be less negative.

4.(a) There is more disorder in the products than in the reactants and $\Delta S°$ would be expected to be positive.

(b) Decrease. See answer to 3(b).

5.(a) $\Delta G° = \Delta H° - T\Delta S° = + 89.5 \text{ kcal mole}^{-1}$

$$K = 2.27 \times 10^{-66}$$

(b) No. $\Delta G°$ positive.

(c) For $\Delta G° = 0$, $\Delta H° = T\Delta S°$.

$$T = \frac{\Delta H°}{\Delta S°} = \frac{104}{0.0485} = 2,144°K$$

Above $2,144°K$ $\Delta G°$ will be negative and the reaction is feasible.

6.(a) $C_2H_6(g)$ is most feasible to prepare under standard conditions.

(b) At first glance one would predict that $C_2H_2(g)$ would be the best fuel on the basis of feasibility of the oxidation reaction. However, balanced equations should be written and this point of view reconsidered.

7. $\Delta G° = \Delta H° - T\Delta S°$

$$= 20.8 - (298) (36 \times 10^{-3})$$

$= 10$ kcal mole^{-1}.

The reaction is feasible at a higher temperature where $\Delta G°$ becomes negative.

Chapter 15.

1.

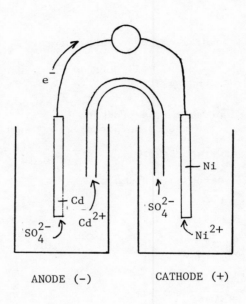

ANODE (−)   CATHODE (+)

2.

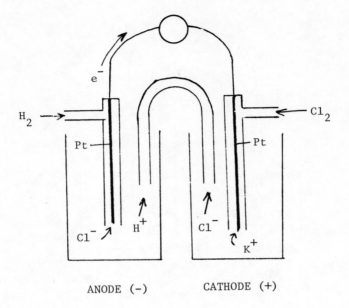

ANODE (−)   CATHODE (+)

3.

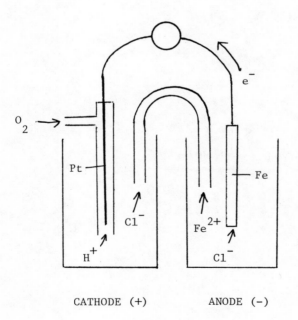

CATHODE (+)   ANODE (−)

4. (a)

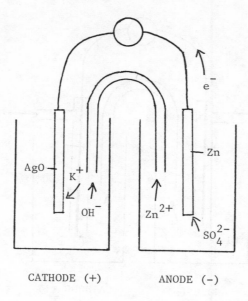

CATHODE (+)   ANODE (−)

(b) $2AgO + H_2O + 2e^- \rightarrow Ag_2O + 2OH^-$

$Zn + 2OH^- \rightarrow Zn(OH)_2 + 2e^-$

———————————————————————————

$2AgO + Zn + H_2O \rightarrow Ag_2O + Zn(OH)_2$

$E = + 0.57 + 1.25$

$= 1.82 \text{ v.}$

(c)

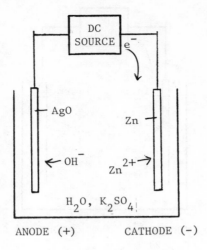

ANODE (+)   CATHODE (−)

5.(a) $Mn(s) \rightarrow Mn^{2+}(aq) + 2e^-$ $E_1^\circ = + 1.18v$

     $\underline{2H^+ + 2e^- \rightarrow H_2(g)}$ $E_2^\circ = \ \ 0.00$

     $Mn(s) + 2H^+ \rightarrow Mn^{2+}(aq) + H_2(g)$ $E^\circ = + 1.18v$

     Reaction will occur.

 (b) $Ag^+(aq) + e^- \rightarrow Ag(s)$ $E_1^\circ = + 0.80v$

     $2Br^-(aq) \rightarrow Br_2 + 2e^-$ $E_2^\circ = \ \ 1.087v$

     No reaction.

 (c) Reaction will occur.

 (d) Reaction will occur.

6. $\Delta G^\circ = -62.7$ kcal mole^{-1}

      $K = 9.71 \times 10^{45}$

7. $[H^+] = 5.4 \times 10^{-2}$.

8. (a)

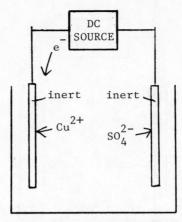

CATHODE (−)   ANODE (+)

$$Cu^{2+} + 2e^- \rightarrow Cu(s)$$

$$2H_2O \rightarrow O_2(g) + 4H^+ + 4e^-$$

8. (b)

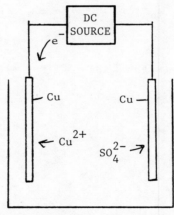

CATHODE (−)   ANODE (+)

$$Cu^{2+} + 2e^- \rightarrow Cu(s)$$

$$Cu(s) \rightarrow Cu^{2+} + 2e^-$$

CHAPTER 16.

1. The microwave has the higher energy.

2. (a) $E = h\nu = 6.6 \times 10^{-27} \times 3.0 \times 10^6 = 2.0 \times 10^{-20}$ erg.

   (b) $\lambda = \dfrac{1}{\bar{\nu}} = \dfrac{c}{\nu} = \dfrac{3.0 \times 10^{10}}{3.0 \times 10^6} = 1.0 \times 10^4$ cm.

3. (a) $E = 9.9 \times 10^{-9}$ erg.

   (b) $E = 2.0 \times 10^{-10}$ erg.

   (c) $E = 1.3 \times 10^{-12}$ erg.

   (d) $E = 1.3 \times 10^{-18}$ erg.

4. $A = C \times \varepsilon \times \ell$

      $C = \dfrac{A}{\varepsilon \ell} = \dfrac{0.750}{2 \times 4815} = 7.79 \times 10^{-5}$ M.

5. $\varepsilon = \dfrac{A}{C\ell} = \dfrac{0.942}{5.80 \times 10^{-4} \times 5} = 3.25 \times 10^2 M^{-1} cm^{-1}$.

Chapter 17.

1. (a) $^{235}_{92}U \rightarrow {}^{0}_{-1}e + {}^{235}_{93}Np$

   (b) $^{131}_{53}I \rightarrow {}^{131}_{54}Xe + {}^{0}_{-1}e$

   (c) $^{43}_{21}Sc \rightarrow {}^{42}_{20}Ca + {}^{1}_{1}H$

   (d) $^{38}_{19}K \rightarrow {}^{38}_{18}Ar + {}^{0}_{1}e$

   (e) $^{197}_{80}Hg + {}^{0}_{-1}e \rightarrow {}^{197}_{79}Au$

   (f) $^{238}_{92}U \rightarrow {}^{4}_{2}He + {}^{234}_{90}Th$

   (g) $^{253}_{99}Es + {}^{4}_{2}He \rightarrow {}^{256}_{101}Md + {}^{1}_{0}n$

(h) $^{10}_{5}B$ + $^{1}_{0}n$ → $^{11}_{5}B$ + γ

2. (a) $^{85}_{34}Se$

   (b) $^{99}_{40}Zr$

   (c) $^{139}_{54}Xe$

   (d) $^{4}_{2}He$

3. Mass of 8 protons = 8(1.00728) =   8.05824 amu
    Mass of 10 neutrons = 10(1.00867) = $\underline{10.0867}$ amu
                                        18.1449 amu

    Mass defect = 18.1449 − 17.99477
                = 0.1501 amu.

    Binding energy = 0.1501 amu x 931.5 $\frac{MeV}{amu}$ = 139.8 MeV.

    Binding energy per nucleon = $\frac{139.8 \text{ MeV}}{18 \text{ nucleons}}$ = 7.777 $\frac{MeV}{nucleon}$.

4. 8.29 $\frac{MeV}{nucleon}$.

5. Δm = 239.0006 − [238.0003 + 1.00867].

       = −0.0084 amu

    ΔE = −0.0084 amu x 931.5 $\frac{MeV}{amu}$ = −7.8 MeV.

6. Δm = [6(1.00867) + 89.8864 + 143.8816] − [1.00867 + 239.0006]

       = −0.1893 amu

    ΔE = −0.1893 x 931.5 = −176.3 MeV.

7. Δm = [1.00867 + 4.00150] − [3.01550 + 2.01355]

       = −0.01888 amu

    ΔE = −17.59 MeV.

370

8. (a) $2\,{}^{2}_{1}\mathrm{H} \;\rightarrow\; {}^{4}_{2}\mathrm{He}$; $\Delta H = -5.962$ MeV

   (b) ${}^{235}_{92}\mathrm{U} + {}^{1}_{0}\mathrm{n} \;\rightarrow\; {}^{90}_{38}\mathrm{Sr} + {}^{144}_{58}\mathrm{Ce} + 2\,{}^{1}_{0}\mathrm{n} + 4\,{}^{0}_{-1}\mathrm{e}$

   $\Delta H = -199.8$ MeV.

Chapter 18.

1.   See text.

2.   rate = k [A] [B]
      $k = 928\ \mathrm{M}^{-1}\mathrm{s}^{-1}$

3.   See Section 18-6.

4.   See text.

5. (a) Test equations 18-10, 18-11, and 18-12.
      The reaction is second order.

   (b) $0.011\ \mathrm{M}^{-1}\mathrm{s}^{-1}$.

6.   For a first order reaction the half-time is independent of
      concentration whereas for a second order reaction the half-
      time depends on inverse concentration.

7.   Try various plots of the data according to the equations
      in Table 18-1. This reaction is first order with $k = 0.051\mathrm{s}^{-1}$.

8.   This reaction is second order with $k = 0.00944\ \mathrm{M}^{-1}\mathrm{s}^{-1}$.

9. $E_a = 21.9\ \mathrm{kcal\ mole}^{-1}$.

Chapter 19

1. (a) +3

   (b) +2

   (c) +3

   (d) +3

   (e) +2

2. (a)

$$\begin{bmatrix} NH_3 \\ H_3N \quad NH_3 \\ Co \\ NH_3 \quad NH_3 \\ NH_3 \end{bmatrix}^{3+}, \qquad \begin{bmatrix} Cl \\ H_3N \quad NH_3 \\ Co \\ H_3N \quad NH_3 \\ NH_3 \end{bmatrix}^{2+},$$

$$\begin{bmatrix} Cl \\ H_3N \quad Cl \\ Co \\ H_3N \quad NH_3 \\ NH_3 \end{bmatrix}^{1+}, \qquad \begin{bmatrix} Cl \\ H_3N \quad NH_3 \\ Co \\ H_3N \quad NH_3 \\ Cl \end{bmatrix}^{1+}.$$

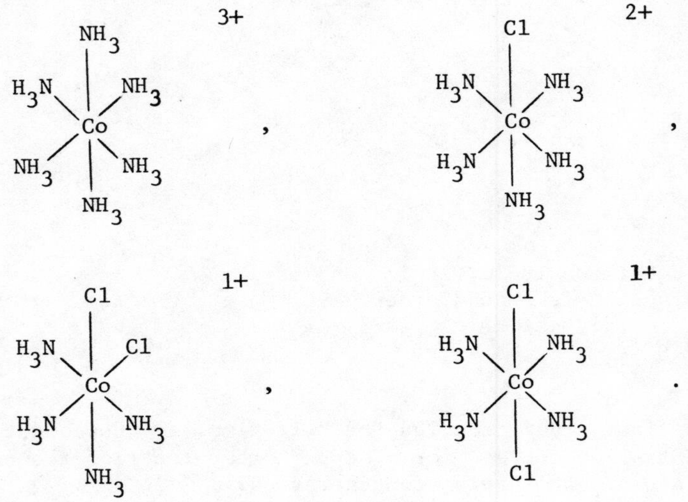

   (b) They exist as <u>cis</u>- and <u>trans</u>-isomers as shown above.

3. $Ni(CN)_6^{2-}$.

4. $Ti(CN)_6^{3-}$ has the largest Δ. $Ti(CN)_6^{3-}$ might be expected to be yellow-orange while $TiCl_6^{3-}$ might be expected to be blue-green.

5.

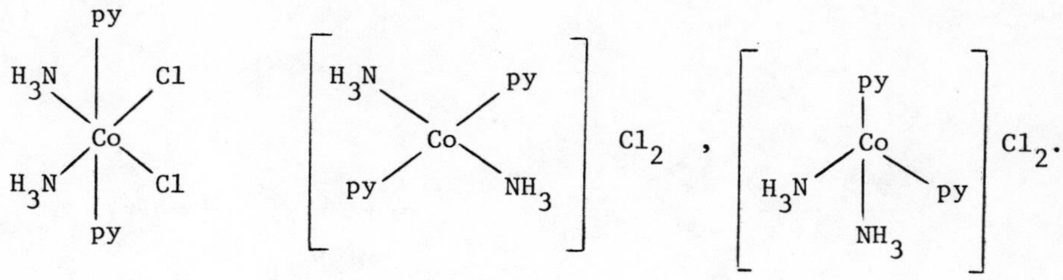

py
H₃N — Co — Cl
H₃N — Co — Cl
py

Co center with py (top), Cl (right), py (bottom), H_3N (left positions), two Cl.

[H₃N, py, py, NH₃ around Co] Cl₂ , [py, H₃N, py, NH₃ around Co] Cl₂.

cis or trans   cis or trans

One could differentiate between the octahedral case and the
others since an octahedral complex would theoretically give no
ionic chloride. The square planar case has two isomers whereas
the tetrahedral has only one.

6.(a)

↑_ ↑_   __ __

↑↓_ ↑_ ↑_   ↑↓_ ↑↓_ ↑↓_

$[Fe(OH_2)_6]^{2+}$ $[Fe(CN)_6]^{4-}$

(b)

__ __   ↑_ __

↑_ ↑_ ↑_   ↑_ ↑_ ↑_

$[Cr(OH_2)_6]^{3+}$ $[Cr(OH_2)_6]^{2+}$

373

(c)

$\underline{\quad}$ $\underline{\quad}$ $\underline{\uparrow}$ $\underline{\quad}$

$\underline{\uparrow\downarrow}$ $\underline{\uparrow\downarrow}$ $\underline{\uparrow\downarrow}$ $\underline{\uparrow}$ $\underline{\uparrow}$ $\underline{\uparrow}$

$[Co(NO_2)_6]^{3-}$ $[Mn(OH_2)_6]^{3+}$

7. (a) Zero.

   (b) Two for tetrahedral and zero for square planar.

   (c) Three for tetrahedral.

8. (a)

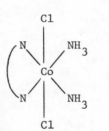 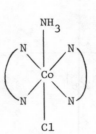

   (b)

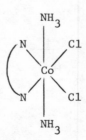

374

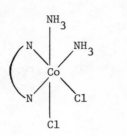

Chapter 20.

1.(a) $Br_2(\ell)$ + $2NaI(s)$ → $2NaBr(s)$ + $I_2(s)$

(b) No reaction.

(c) $NaCl(s)$ + $H_3PO_4(\ell)$ → $HCl(g)$ + $NaH_2PO_4(s)$

(d) $2ZnS(s)$ + $3O_2(g)$ → $2ZnO(s)$ + $2SO_2(g)$

(e) $P_4O_{10}(g)$ + $10C(s)$ → $P_4(g)$ + $10CO(g)$

(f) $AsCl_3(\ell)$ + $3H_2O$ → $H_3AsO_3(\ell)$ + $3H^+(aq)$ + $3Cl^-(aq)$

(g) $NaNO_3(s)$ + $H_2SO_4(\ell)$ → $NaHSO_4(s)$ + $HNO_3(g)$

(h) $Cl_2(g)$ + $2OH^-(aq)$ → $Cl^-(aq)$ + $ClO^-(aq)$ + H_2O

(i) $As_2O_3(s)$ + $3C(s)$ → $2As(\ell)$ + $3CO(g)$.

2. HNO_2, NO, N_2O.

3.   See equations 20-30, 20-31, and 20-32 in text.
The NO_2 is both reduced and oxidized in the same reaction
(20-32).

4.(a) See equation 20-24 in text.

(b) $P_4O_6(s)$ + $6H_2O$ → $4H_3PO_4(aq)$.

(c) $Sb_2O_3(s)$ + $H_2(g)$ → $2Sb(s)$ + $3H_2O$.

(d) $NH_4NO_2(s)$ → $N_2(g)$ + $2H_2O$.

5.(a) Yes.

(b) No.

(c) The immediate products are S_8 and SO_2 with SO_2 disproportionating further to SO_4^{2-}.

6. P_4 is not stable. As is stable.

Chapter 21.

1.
$$C - C - C - C - C - C - C$$

n - heptane

$$\begin{array}{c} C \\ | \\ C - C - C - C - C - C \end{array}$$

2 - methylhexane

$$\begin{array}{c} C - C - C - C - C - C \\ | \\ C \end{array}$$

3 - methylhexane

$$\begin{array}{c} C \\ | \\ C - C - C - C - C \\ | \\ C \end{array}$$

2,2-dimethylpentane

$$\begin{array}{c} C \\ | \\ C - C - C - C - C \\ | \\ C \end{array}$$

3,3-dimethylpentane

$$\begin{array}{c} C - C - C - C - C \\ \quad | \quad | \\ \quad C \quad C \end{array}$$

2,3-dimethylpentane

$$\begin{array}{c} C - C - C - C - C \\ | \quad\quad | \\ C \quad\quad C \end{array}$$

2,4-dimethylpentane

$$\begin{array}{c} C \\ | \\ C - C - C - C \\ | \quad | \\ C \quad C \end{array}$$

2,2,3-trimethylbutane

$$\begin{array}{c} C - C - C - C - C \\ | \\ C \\ | \\ C \end{array}$$

3-ethylpentane.

2.   See sections 21-1 and 21-2.

3.(a)
```
              C
              |
    C - C - C - C - C
              |
              C
              |
              C
```

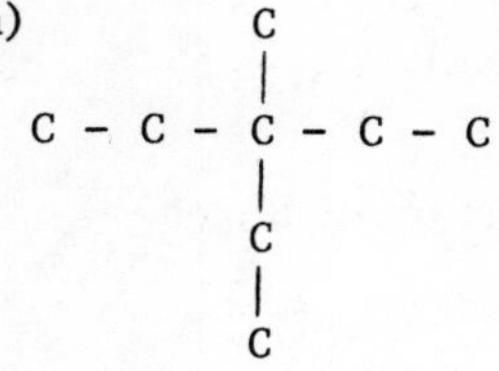

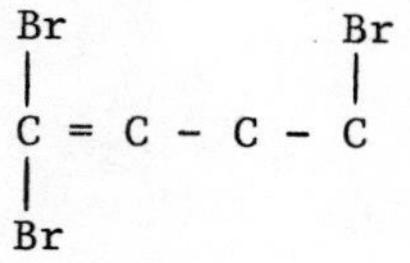

(b)
```
                  C
                  |
                  C
                  |
    C - C - C - C - C - C
        |   |       |
        C   C       C
```

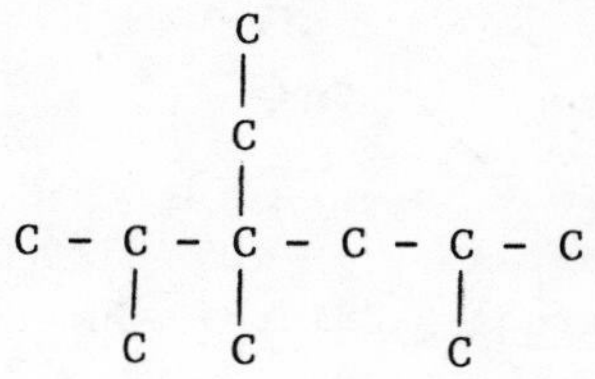

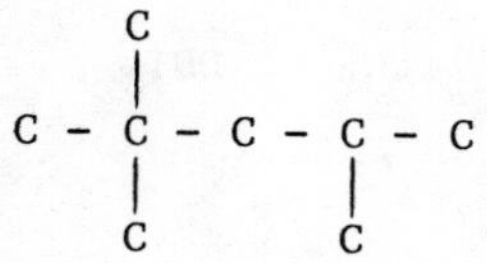

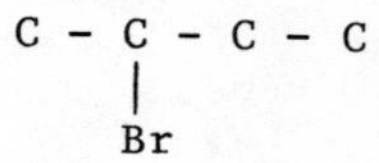

(c)
```
          C
          |
    C - C - C - C - C
        |       |
        C       C
```

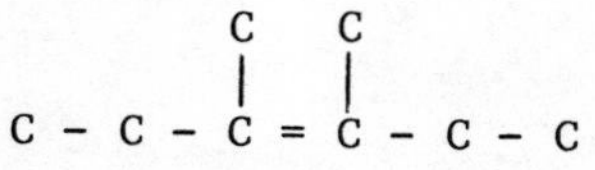

(d)
```
    C - C - C - C
        |
        Br
```

4.(a)
```
    Br          Br
    |           |
    C = C - C - C
    |
    Br
```

(b)   C - C = C - C - C

(c)
```
          C   C
          |   |
    C - C - C = C - C - C
```

(d)
```
          C                   C
          |                   |
    C = C - C - C - C - C - C - C
```

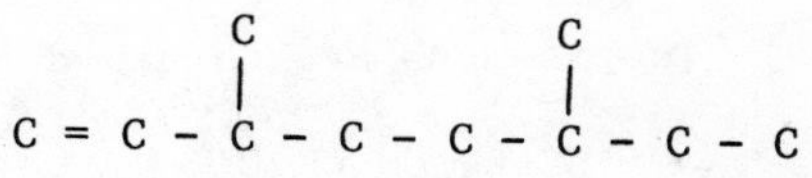

5.(a)

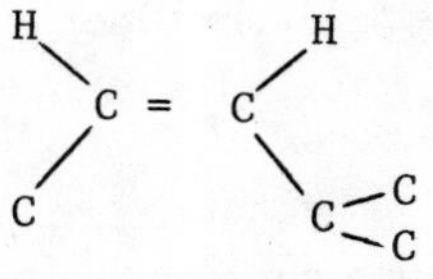

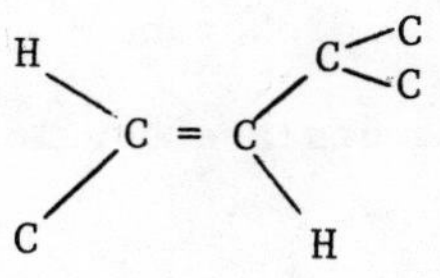

cis-4-Methyl-2-pentene   trans-4-Methyl-2-pentene

(b)   No isomers possible.

(c)   No isomers possible.

(d)

   trans-2-Chloro-2-butene   cis-2-Chloro-2-butene.

6.(a) 3-Methyl-1-butyne

  (b) 2-Hydroxy-3-chlorophenol

  (c) 1,1,1-Trichloro-2,2-bis-(p-chlorophenyl)ethane, DDT.

  (d) 2,6-Dinitrotoluene.

7.(a) One product is obtained: CH_3CH_2Cl, chloroethane.

  (b) Two isomers are obtained:

$$CH_3CH_2CH_2Cl \quad \text{and} \quad CH_3\underset{\underset{Cl}{|}}{C}HCH_3$$

       1-Chloropropane   2-Chloropropane

  (c) Two isomers are obtained:

$$CH_3CH_2CH_2CH_2Cl \quad \text{and} \quad CH_3CH_2\underset{\underset{Cl}{|}}{C}HCH_3$$

       1-Chlorobutane   2-Chlorobutane

  (d) Two isomers are obtained:

$$CH_3\underset{\underset{CH_3}{|}}{C}HCH_2Cl \quad \text{and} \quad CH_3\underset{\underset{Cl}{|}}{\overset{\overset{CH_3}{|}}{C}}CH_3$$

    1-Chloro-2-methylpropane   2-Chloro-2-methylpropane

8. $C_8H_{18}(\ell) + 12\frac{1}{2}O_2(g) \rightarrow 8CO_2(g) + 9H_2O(g)$.

378

9.(a)

$$CH_3 - \underset{\underset{Br}{|}}{\overset{\overset{CH_3}{|}}{C}} - \underset{\underset{Br}{|}}{CH_2}$$

(b) CH_3CHICl

(c)

$$CH_3 - \underset{\underset{H}{|}}{\overset{\overset{CH_3}{|}}{C}} - CH_3Br$$

(d) $\xrightarrow{\underline{HCl}} CH_3C=CH_2 \xrightarrow{\underline{HI}} CH_3\overset{\overset{I}{|}}{\underset{\underset{Cl}{|}}{C}}CH_3$

$\qquad\qquad\quad \underset{\underset{Cl}{|}}{} $

(e) $\xrightarrow{\underline{Br_2}} CH_3\underset{\underset{Br}{|}}{\overset{}{C}} = \underset{\underset{Br}{|}}{CH} \xrightarrow{\underline{Br_2}} CH_3\underset{\underset{Br}{|}}{\overset{\overset{Br}{|}}{C}} - \underset{\underset{Br}{|}}{\overset{\overset{Br}{|}}{CH}}\;.$

10.(a)

$$CH_3 - \underset{\underset{I}{|}}{\overset{\overset{CH_3}{|}}{C}} - CH_2 - CH_3$$

2-Iodo-2-methylbutane

(b)

$$CH_3 - \underset{\underset{O\,S\,O_3H}{|}}{\overset{\overset{CH_3}{|}}{C}} - CH_3$$

tert-Butyl hydrogen sulfate

(c)

$$CH_3 - \underset{\underset{H}{|}}{\overset{\overset{CH_3}{|}}{C}} - CH_2Br$$

Isobutyl bromide

379

(d)

$$CH_3 - \underset{\underset{\displaystyle CH_3}{|}}{CH}CH_3$$

2-Methylpropane

(e) $(- CH_2C(CH_3)_2 -)$

Butyl rubber

11. (a) Nitrobenzene

(b) m-Bromonitrobenzene

(c) o- and p-Nitrotoluene.

(d)

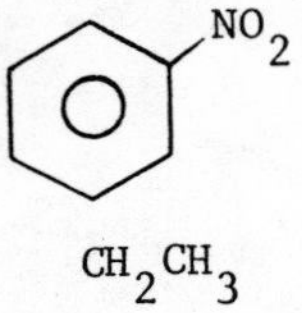

(e)

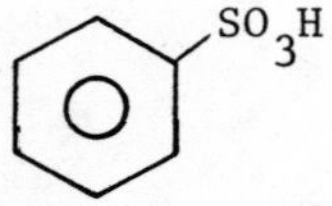

Chapter 22.

1. (i) $Na^+H^- + CH_3CH_2OH \rightarrow Na^+OCH_2CH_3^- + H_2$

(ii) CH_3CH_2I

(iii) $CH_3CHClCH_3$.

(iv)

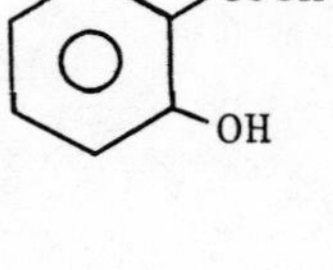

$+ CH_3COOH$

(v) $CH_3C = O$ + $2NH_3$ → $CH_3C = O$ + NH_4^+ Cl^-

with Cl below the left carbon, and NH_2 below the right carbon.

$$\underset{\underset{Cl}{|}}{CH_3C} = O \ + \ 2NH_3 \ \rightarrow \ \underset{\underset{NH_2}{|}}{CH_3C} = O \ + \ NH_4^+ \ Cl^-$$

(vi)   <u>aldehydes</u> and <u>ketones</u>

(vii) $RCOOH + PCl_5 \rightarrow RCOCl + HCl + POCl_3$

         $RCOOH + NH_3 \rightarrow RCONH_2 + H_2O$

(viii)   The functional group families are aldehyde, ketone, and alcohol with structures

$$\underset{}{\overset{O}{\overset{\|}{-C}}-H,} \qquad \overset{O}{\overset{\|}{-C}-,} \qquad and \qquad -OH.$$

2.(i)   carboxylic acids

 (ii)   aldehydes

 (iii)   esters

 (iv)   aldehydes (acids)

 (v)   addition

 (vi) $\underset{\underset{H}{|}}{CH_3-C}=NNHCONH_2$

 (vii)

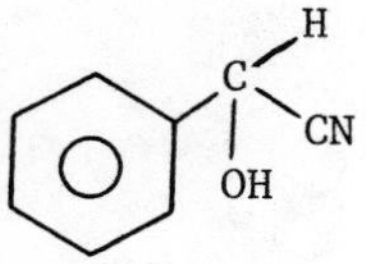

(viii) $2Ag +$ COO^- + $4NH_3$ + H_2O

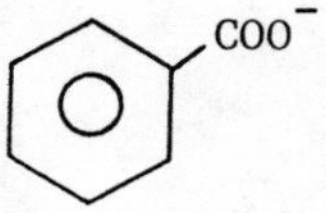

          silver mirror

 (ix)   No reaction.

3.(i)   (a) Two, (b) These are enantiomers.

  (ii)   (a) None,   (b) None.

  (iii)   (a) None, (b) None.

  (iv)   (a) Two, (b) These are enantiomers.

  (v)   (a) Two stereoisomers, (b) None.

  (vi)   (a) Three stereoisomers, (b) None.

4.   A planar complex cannot be optically active.

5. (i)   Four stereoisomers, four optical isomers.

  (ii)   Four stereoisomers, four optical isomers.

  (iii)   Three stereoisomers, two optical isomers.

  (iv)   Two stereoisomers, two optical isomers.

6.   See section 22-6 in text.

Chapter 23.

1.

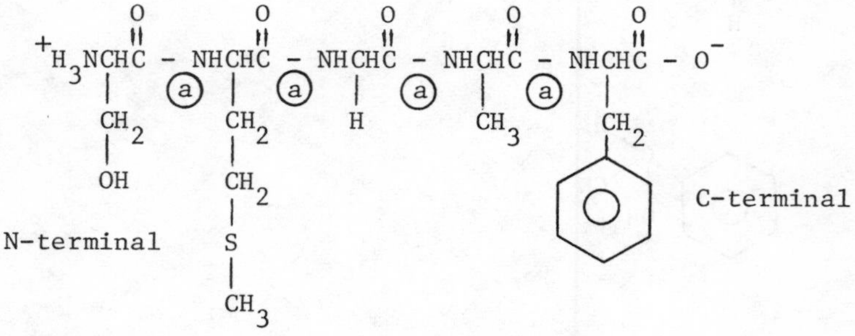

     (a) = peptide bonds

2.(a)   val – arg – phe – tyr – cys – cys

           or

        cys – val – arg – phe – tyr – cys

(b)   React the hexapeptide with an N-terminal specific reagent and identify the N-terminal amino acid as val or cys.

(c)   See answer 1. above.

3.(a)

$$O_2N-\bigcirc-NH-CH\underset{|}{C}-NH-\underset{|}{CH}-COO^-$$

with the structure:
$O_2N-\text{(ring)}-NH-CHC(=O)-NH-CH-COO^-$, ring substituted with NO_2, the CH bearing CH_2-SH, and the CH bearing CH_3.

(b)

$$O_2N-\bigcirc-NH-CH-\overset{O}{\underset{||}{C}}-OH \;+\; {}^+H_3N-CHCOOH$$

ring substituted with NO_2, the CH bearing CH_2-SH, and the right CH bearing CH_3.

(c), (d)   Follow same procedure as (a), (b) above.

4.   Reactions characteristic of the -COOH group.

5.(a) $C_6H_{12}O_6(s) + 6O_2(g) \rightarrow 6CO_2(g) + 6H_2O(\ell)$

(b)   Synthesis and heat.

(c)   ATP $\rightarrow$ ADP reaction is coupled to other reactions.

6.   See text.

7.(a) $CH_3OH \rightarrow HC\overset{O}{\underset{H}{\diagdown}} \rightarrow HCOOH$

(b)   No. Both products are toxic.

(c)   Drink "martinis" so as to allow the detoxification reaction for CH_3CH_2OH to take place instead of reaction (a). While the former reaction is occurring the wood alcohol may be excreted.

383

8.   See text, section 23-8.

9.   See text, section 23-9.